THE RAILWAYS OF PURBECK

by

R.W. Kidner

THE OAKWOOD PRESS

ISBN 0 85361 372 9

Typeset by Gem Publishing Company, Brightwell, Wallingford, Oxfordshire.

Printed by Bocardo Press.

First published 1973
Reprinted 1979
New Enlarged edition 1988

Published by
The OAKWOOD PRESS
P.O.Box 122, Headington, Oxford.

Contents

Acknowledgements

I would like to thank the many people who have assisted in elucidating the facts included in this book, and especially: Mr L. Popplewell, Swanage Library, the Dorset County Library, Mr H.V. Borley, Mr C.R. Legg, Mr R. Wear, Mr D. Gould, Mr J.I.C. Boyd, Mr R. Roberts, Swanage U.D.C. Engineer, Major J. Mansell of Smedmore House, Mr S.M. Barrell, Mr D. Faulkner, Mr G. Brigden, and the Swanage Railway.

Bibliography

The following have been consulted:

Pike Bros, Fayle & Co. Ltd by W.J.K. Davies (Narrow Gauge Railway Society Handbook No. 1)

The Middlebere Tramway by B. Baxter (Railway & Canal Historical Society).

Brownsea's Lost Railway by L. Popplewell (Dorset Yearbook).

Purbeck Island by R. Legg.

Unknown Dorset by D. Maxwell.

Inventory of Historical Monuments in the County of Dorset (Vol. 2, S.E.)

Stephen Lewin and the Poole Foundry by R. Wear & E. Lees (IRS)

The Railway Magazine

The Southern Railway Magazine

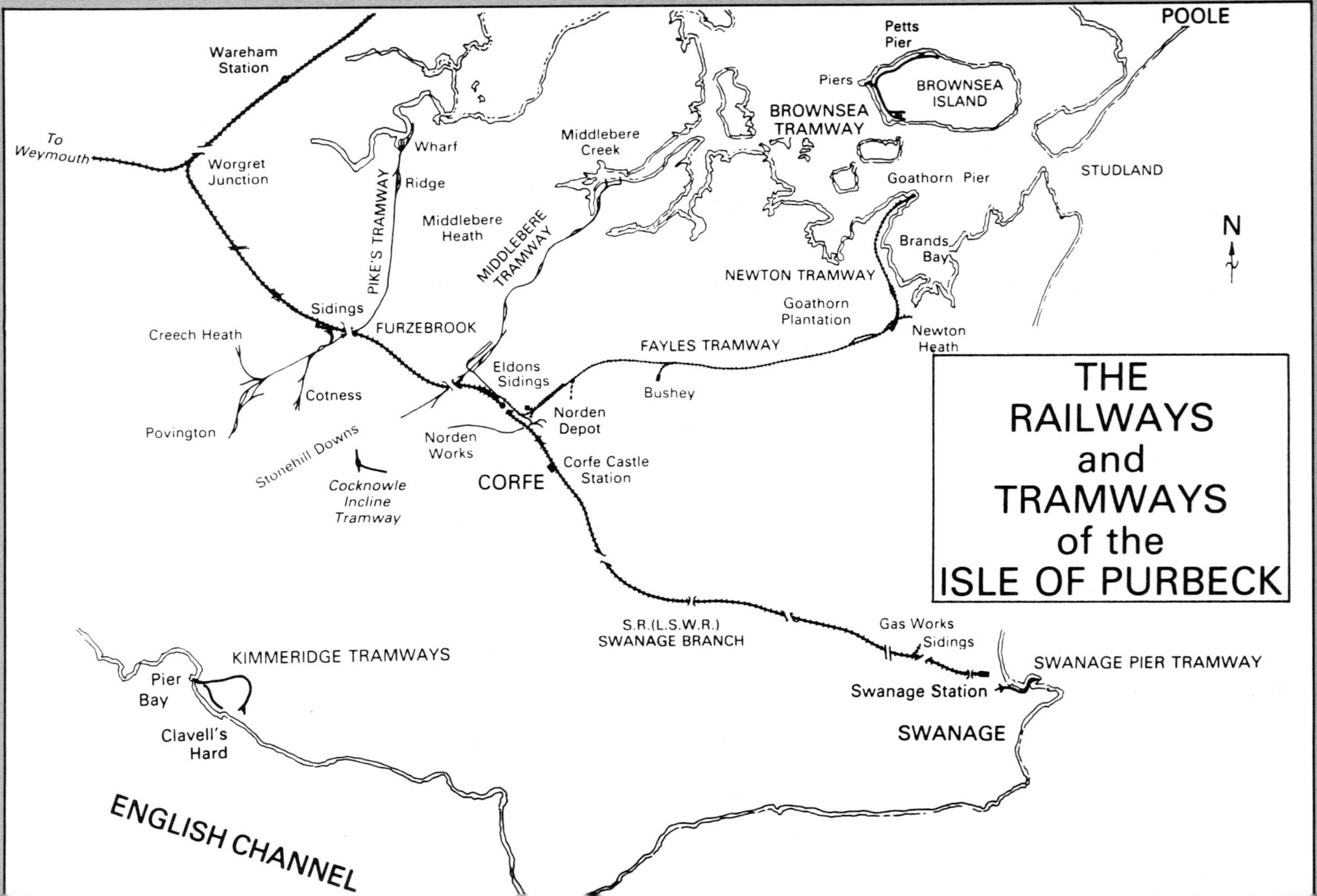
THE RAILWAYS and TRAMWAYS of the ISLE OF PURBECK
POOLE
Petts Pier
Piers
BROWNSEA ISLAND
BROWNSEA TRAMWAY
STUDLAND
Goathorn Pier
N
Brands Bay
NEWTON TRAMWAY
Goathorn Plantation
Newton Heath
FAYLES TRAMWAY
Bushey
Wareham Station
To Weymouth
Worgret Junction
Wharf
Ridge
PIKE'S TRAMWAY
Middlebere Creek
Middlebere Heath
MIDDLEBERE TRAMWAY
Sidings
FURZEBROOK
Creech Heath
Cotness
Povington
Eldons Sidings
Norden Depot
Norden Works
Stonehill Downs
Cocknowle Incline Tramway
CORFE
Corfe Castle Station
S.R.(L.S.W.R.) SWANAGE BRANCH
Gas Works Sidings
SWANAGE PIER TRAMWAY
Swanage Station
SWANAGE
KIMMERIDGE TRAMWAYS
Pier Bay
Clavell's Hard
ENGLISH CHANNEL

Introduction

The "Isle" of Purbeck comprises a promontory extending some ten miles south-west of the River Frome at Wareham, being bounded on the north by the river and on the north-east by the ramifications of Poole Harbour, a mass of winding channels through which in ancient days this area was approached, before other means of communication existed. The "island" is bisected by a ridge of hills running east to west through Corfe, and it is the combination of limestone plateau, clay vale, and chalk ridges which gives this small area such a variety and interest. The clay vale was settled in mesolithic times, certainly before 3500 B.C., and by the twelfth century working of a local stone, and of the scarce material known as "Purbeck Marble" had begun, being used in various cathedrals, including Salisbury and Exeter, and in Westminster Abbey. Stone workings were mostly in the area near North Swanage, Langton, and Worth Matravers, and the stone seems to have been taken along narrow tracks to the nearest navigable creeks in Poole Harbour, and later to Swanage Bay.

The other important mineral in the "island" was ball clay, found in the vale north of the Purbeck Hills. This is used in the pottery industry and is scarce; the Purbeck deposits are said to be superior to the rather better-known ones in Devon and Cornwall. These deposits, and the sand that was found with them were probably worked in mediaeval times, but it was about 1760 that it began to be an industry. As usual in mining, transport was the problem, and the water was the answer. As will be described later, tramways were built to carry the products to Poole Harbour, the earliest one being laid in 1806.

There were several ancient little ports along the southern margin of Poole Harbour, which forms the north shore of Purbeck: Wytch, the port of Corfe Castle, Russell Quay, Middlebere Creek — all were on relatively deep channels leading into the main channel for seaborne traffic running into the port of Poole. That this formed a satisfactory method of transport is proved by the fact that in Purbeck, unlike many other mining areas, the main line railway when it came did not immediately draw on to itself the traffic from the mines.

About a hundred years ago, the holiday "industry" began to assume importance; the fishing village of Swanage with its improvised methods of shipping stone had given way to a small Victorian town with a pier. All the trappings of Victorian holiday-making began to appear– and there can be few seaside towns with more masonic knick-knacks — so that by the end of the 1914–18 War, when family holidays became general and not the perquisite of the upper-middle class, the town was ready to boom.

For the railway-minded holiday-maker, Purbeck had a special charm. The arrival of each train was something of an event, particularly if it was a "special" direct from London, with two locomotives and a good deal too long to fit in the platform. Those with cameras found a composition with a Southern tank engine in the foreground and Corfe Castle on the skyline quite irresistible. And for the very few who in those days recognised industrial railways as being worthy of study, they could take off on foot into the roadless wastes of Wytch Heath, Newton Heath or Stoborough and come upon such sights as they had not dreamed of. When the author first visited the area, nothing had appeared in print concerning the clay tramways, and the first sight of the Lewin engine on Fayles Tramway, and of the evening gravity wagon for Ridge clicking and rocking its way across the heath, had all the excitement of discovery.

Fortunately, the London & South Western Railway tended to keep its seaside termini in their place; not for them the great iron structures and acres of excursion sidings that uglified so many Great Northern Railway resorts. Swanage station remained a gem; dignified with its stone, quiet with its polite little tank engines. Nevertheless, the Swanage Branch almost attained the status of a main line, at least in local eyes. A guide-book to Swanage of the late 'twenties says: "From the Midlands and the North, the L.M.S. Railway, by means of the Somerset & Dorset line, runs down via Bath and Poole. There is also the Didcot–Winchester line of the Great Western Railway, connecting at Southampton with the Southern Railway main line from London to Swanage" — an intriguing approach, not quite shared by officials at the other end of the telescope at Waterloo.

The front page of the 1881 Act for the Swanage Railway

ACT

To Incorporate a Company for the Construction of the Swanage Railway and for other purposes.

[*Royal Assent*, 18*th July*, 1881.]

WHEREAS the construction of Railways to connect Swanage with the London and South Western Railway near Wareham all in the county of Dorset would be of public and local advantage: Preamble.

And whereas the persons in this Act named are willing with others at their own expense to construct such Railways and are desirous of being incorporated into a Company for the purpose:

And whereas plans and sections of the said Railways showing the lines and levels thereof with a book of reference to the plans containing the names of the owners and lessees or reputed owners and lessees and of the occupiers of the lands through which the

Chapter One

The Swanage Railway

In the years immediately following Queen Victoria's accession the targets aimed at by the newly–born railways were limited. It was a new-fangled thing, which had not yet proved itself. On the south coast, there were three places which three separate companies set out to reach: Dover, Brighton and Southampton. The first train from London to Southampton ran in 1840, a year ahead of the Brighton line and four years ahead of Dover. Further west there lay whole counties waiting to be opened up. The Great Western's great sweep to the West was making progress, but well to the north, and when the London & Southampton Railway became the London & South Western, the opening up of Hampshire and Dorset was very much in its mind. The Southampton & Dorchester Railway, promoted locally, came under the wing of LSWR, and was opened on 1st June, 1847, though it was not available to London trains until 29th July, as junctions at Southampton were uncompleted.

A wayside station on this line was set up at Wareham, well outside the town, as the citizens had followed the lead of the New Forest Commissioners in refusing to have any actual contact with the sordid reality of railways. The line was then single, and remained something of a backwater as the LSWR strove to reach the far west by another route, through Salisbury. However, in 1857 the LSWR ran into Weymouth (over Great Western metals), and during 1863 the line from Wimborne through to Dorchester was doubled.

Various attempts were made to promote a branch line from Wareham either to Swanage or to the clay workings. One line, the Isle of Purbeck Railway, succeeded in getting an Act in 1863, but backed down in face of local hostility. However, the town of Swanage was growing rapidly, and could not indefinitely be without modern transport. When the Swanage Railway was promoted and received its Act (44 –45 Vic.c.159)* in 1881, the moment of truth had arrived. The problem for the Wareham worthies was a real one; because they had insisted on the siting of Wareham station north of the town, which had now grown out to it, a direct branch to Swanage would cut the town in half. A Solomon's judgement was produced; the branch would start at Worgret, 1¼ miles to the west, and not pass through Wareham at all. Worgret was a lonely and remote cluster of houses, which would not have had much say in the matter, and did not even rate a station at the junction. On the other hand, Wareham station was too cramped to handle the branch traffic, and plans were drawn up for a larger station on the other side of the level crossing. And so for the whole of its life the Swanage Branch was worked as if it started at Wareham, but its trains ran the first 1¼ miles on the main line to Weymouth.

The railway duly opened on 20th May, 1885. From its inception it was worked by the LSWR, and by an Act of 1886 (49–50 Vic.c.110) the Swanage Railway was absorbed into the larger company.

A description of the line would be fitting here. As stated, the branch left the Weymouth line at Worgret; here there was a passing loop south of the

* The original Act had also given powers for a railway or tramway from Swanage station to the stone "bankers" on the front, and this is described later.

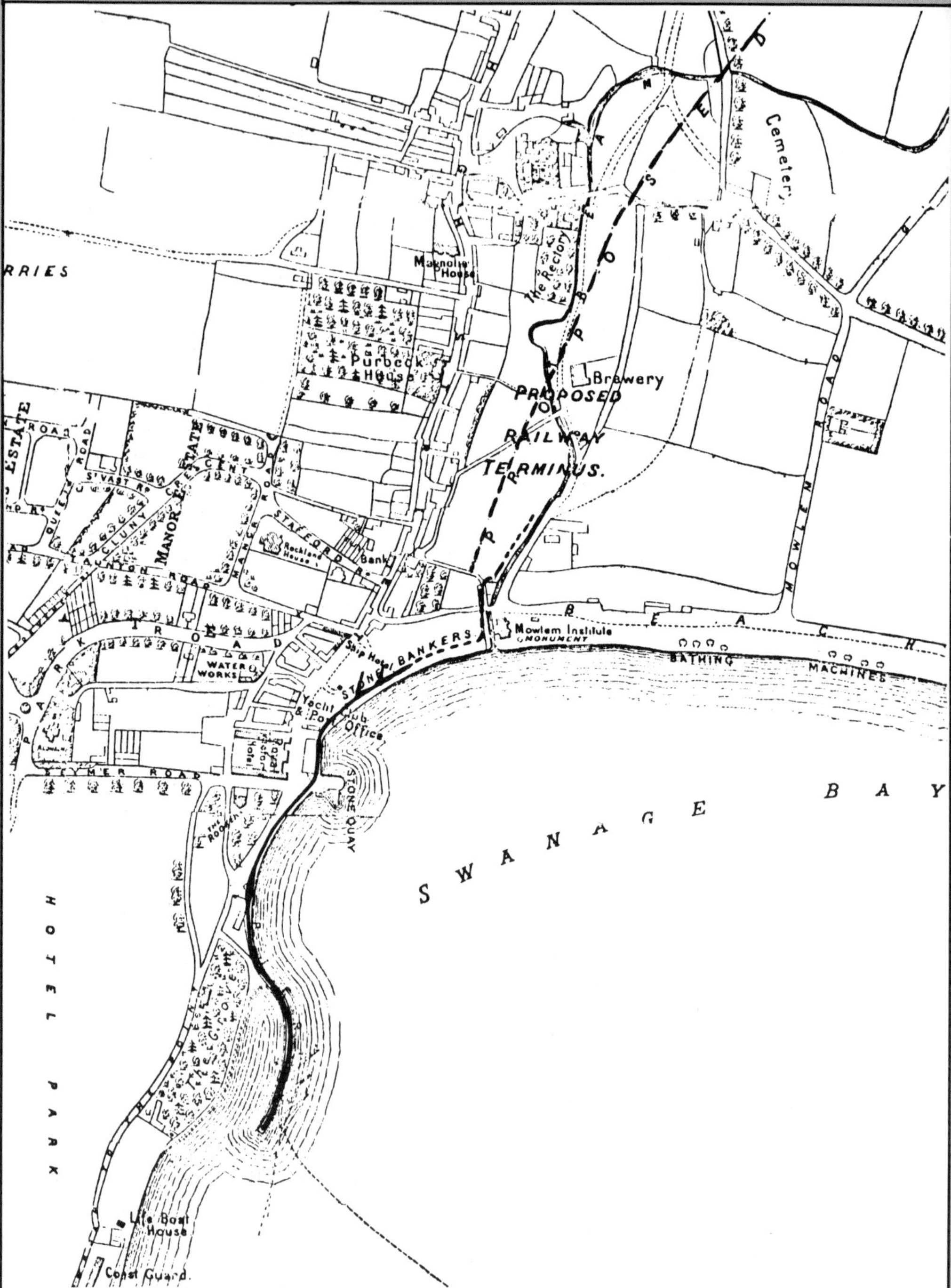

This map of Swanage shows the position of the proposed LSWR station. A dotted line representing a Tramway has been added from the Pier Tramway to the new station; this was not built, but a standard gauge LSWR tramway was built on somewhat the same line. The terminus (as built) did not extend quite so far towards the front as shown here.

junction, so that a train from Swanage could await a path on the up line to Wareham without fouling a down branch train. Worgret Junction box (126 miles from Waterloo) was a busy place in the summer, and also hard work. When in the 'thirties the *Southern Railway Magazine* was trying to find the longest signalman's pull, Worgret claimed on three counts: Swanage Branch up distant 1406 yds, down main distant 1532 yds, up main distant 1680 yds — it did not win however. The branch became single on leaving the loop, crossing the streams of the Frome and various bits of marsh, to run over Grange and Creech Heaths. The latter was already occupied by the Pike Brothers' clay workings, and the LSWR crossed over their line from Ridge near Furzebrook House; no exchange siding seems to have been put in at that time. A mile and a quarter further on, passing under the Wareham–Corfe Road at milepost 130, the Branch met the ancient Middlebere Plateway (Fayle's), and made a narrow tunnel through its embankment to accommodate it, though it seems the pits west of the line may have been more or less worked out. The new line from the Middlebere Plateway to pits in the east ran parallel with the LSWR. Here at a later date the single-line Eldon siding was placed for clay transfer, the name being that of a local landowner. It is not absolutely certain that it was not put in at opening, as the relevant OS map is ambiguous. A short spur from the new Fayle's line crossed the LSWR either at opening or shortly after.

Continuing towards Corfe, it becomes clear that there was no choice of route; it had to be between the high hill on which the castle stands and the equally high East Hill. The road was already there, and the branch, after passing over a magnificent four-arch stone viaduct, had to do some cutting before entering Corfe Castle station (131 m). This is in excellent taste and fits well with the castle and other stone buildings of the village. There was a passing loop and a long and a short siding. The heath now begins again, as the line finds it way down to sea level, with gradients as steep as 1 in 77. It entered the town, when first built, on the north side of what little development had taken place (it was now running west to east, rather than south), and little demolition can have been required. There was a siding with points facing Wareham 150 yds west of the Victoria Avenue bridge, to serve the gasworks. As the branch completed its last steep down gradient, a two-line siding protected by a gate, for stone traffic, could be seen on the right, though this is unlikely to have been there at the time of opening. The 136¼ m. post was on the station platform.

The town the LSWR was arriving to serve was still small, the development being only from the Royal Victoria Hotel up to Northbrook Road. East of Mowlem Institute there was nothing but some stone bankers not served by the pier tramway, and a few bathing machines at the end of Victoria Avenue (then called Mowlem Road). However, growth was rapid for the next 20 years when it largely ceased until after the last war. The Ward Lock guide for 1934 put it well — "a dear inconsequential little town, with all sorts of lovable absurdities in the way of lamp posts, clock towers . . . bestowed during the nineteenth century by citizens who loved it . . .".

Although the station building at Swanage is neat and small, the layout was unusually large. A trailing point led into an attractive one-engine locomo-

CORFE CASTLE
c.1950

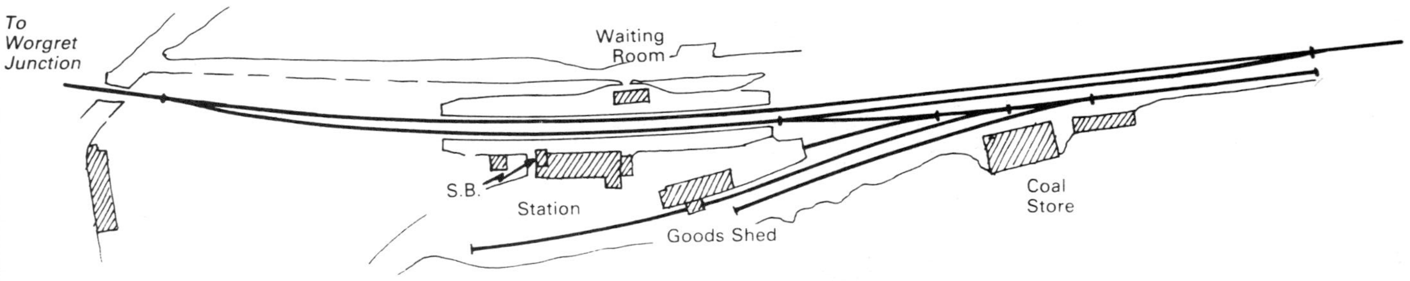

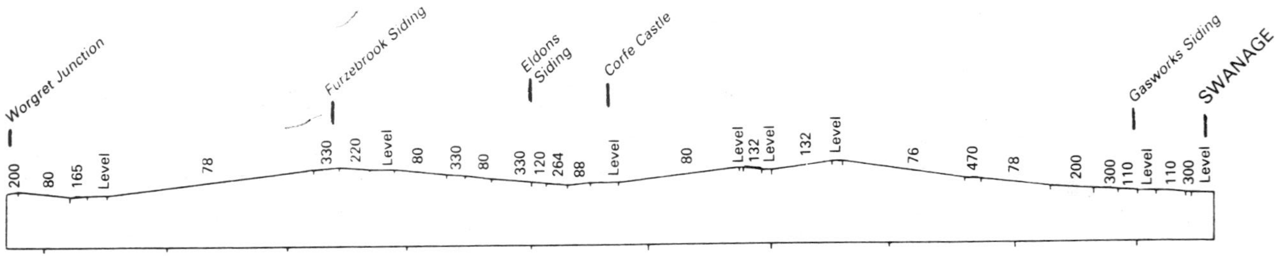

SWANAGE BRANCH GRADIENT PROFILE

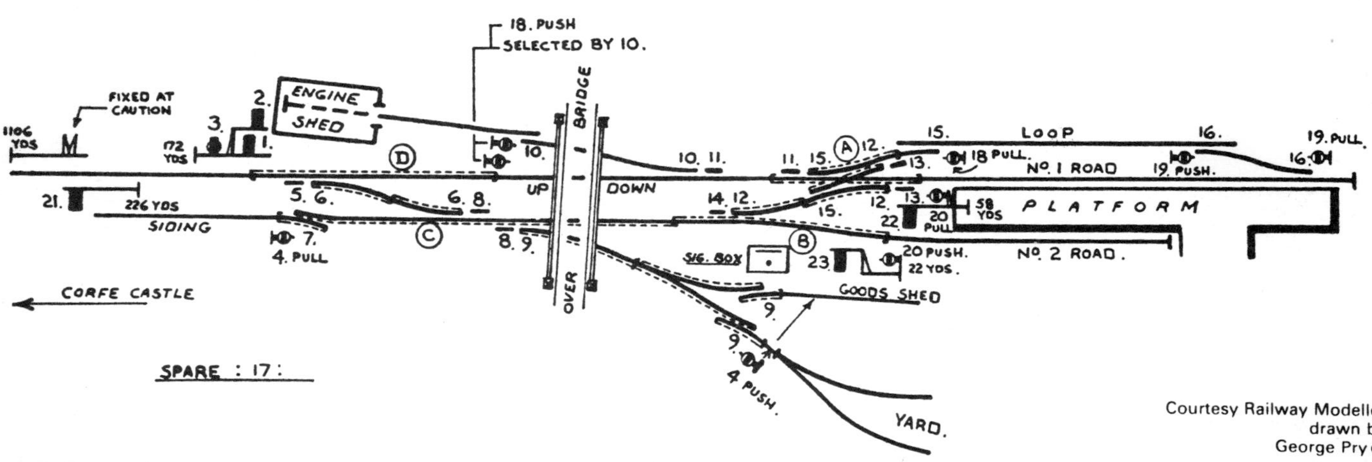

Courtesy Railway Modeller
drawn by
George Pryer

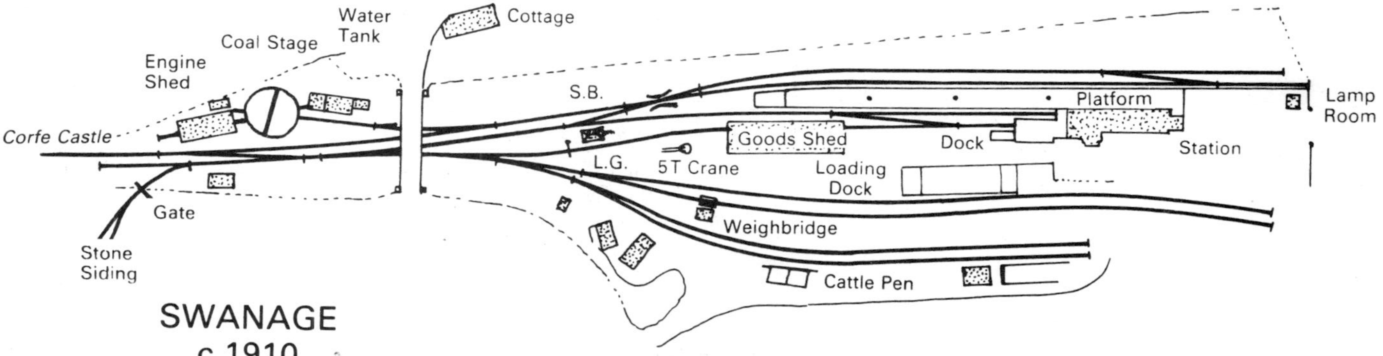

SWANAGE
c.1910

tive shed, with turntable. Then another long quadruple siding ran off to the right before the line passed under a road bridge to enter the station proper. The locomotive shed was built at an angle which made it impossible to run from the station straight across the table and into the shed — so some muscle was required in the morning even for a tank engine. The station had two platforms, that on the south side being even shorter than the main one. It was about equidistant from the Parade and the High Street and therefore well-sited. All mileposts on the branch were numbered from Waterloo; bridge numbers ran from 1 near Worgret box to 31, the overbridge by the turntable at Swanage.

Originally one of the sidings to the south of the station was extended by a tramway running in a narrow alley behind the shops in Station Road, to a turntable on a cross-siding carrying a crane and serving a stone store. This was very close to the stone store on the Pier Tramway on the south side of Institute Road, but of course the lines were of differing gauge. The LSWR tramway was still on the 1906 OS map, but must have closed about that time due to road development south of the station. It seems a fair assumption that it was at this time that the two stone sidings west of the station, mentioned previously, were laid. These were shown in the 1938 *Railway Clearing House Handbook* as belonging to W. Haysom & Sons and Swanworth Quarries Ltd.

The service at opening was modest and inexpensive. The line was worked on the "one engine in steam" basis; the engine left Swanage with the first train at 7.20 am; there were five trains each way and a midday goods; the last train left Wareham at 9.10 pm and the engine returned to its shed. From 1st August, 1885 the goods working was a "mixed" train (i.e. carrying passengers if required); it took 45 minutes. It must be remembered that there was no clay wagon pick-up work to do as Eldon and Furzebrook sidings did not exist, so this timing, which was 18 minutes longer than the passenger timing, only took into account shunting duties at Corfe, and probably an allowance for a bit of crawling up the banks if there was some heavy coal or stone in the train. There was no Sunday service.

Over the years the service was somewhat improved, and through workings in the summer period increased. It is not possible to say exactly when through coaches from Waterloo commenced, as Bradshaw did not make a clear enough distinction between through coaches and connections. In 1922 the "Pines Express", which travelled via the S&DJR, carried a through coach from Manchester to Swanage. The branch train in the down bay could leave as soon as the Weymouth train had cleared the section to Worgret; if there was a through coach the gap in the timetable might be longer, as the branch train would have to back in to the through platform to pick it up. In 1931 (Winter service) there were thirteen down trains on the branch; that leaving Waterloo at 12.30 pm and the 4.30 pm and 6.30 pm had good Swanage connections or through coaches; the 4.30 pm from Bournemouth Central ran through to Swanage, with no connection to Weymouth at Wareham. The connection from the 4.30 ex-Waterloo left Wareham at 7.17 and was followed at 7.47 by a connection off the Bournemouth West to Dorchester train — there were examples in various years of two trains following each other over the branch; the second train passed the first one returning at Corfe

Name of siding.	Position.	(1) Station in charge of working. (2) wagons labelled to †	Gradient at point of connection (1 in)	Catch points provied in sidings at.	Points of siding controlled by or worked from.	If gates provided across siding: Key to be obtained from.	If gates provided across siding: Key to be returned to.	Worked by.	Remarks.
SWANAGE BRANCH.									
Pike's Furzebrook	Up side between Worgret Jct. and Corfe Castle	Corfe Castle	330 falling towards Worgret Jct.	Outgoing ground signal	Ground frame Train Tablet	Corfe Castle		Various goods services	*
Eldon's ...	Down side between Worgret Jct. and Corfe Castle	Corfe Castle	330 falling towards Corfe Castle	Clearance point	Ground frame Train Tablet	Interlocked by siding points		Various goods services	* —
Beavis ...	Up side Swanage	Swanage	270 falling towards Swanage	—	Extension of goods yard	—	—	—	For working instructions see page 45.
Haysom & Sons	Up side Swanage, Corfe Castle end •	Swanage	90 falling towards Corfe Castle	—	Hand points from siding	Secured by bolt from Swanage box		Shunting engine	For working instructions see page 45. *
Swanage Gas Works	Down side between Corfe Castle and Swanage	Swanage	110 falling towards Swanage	In reception siding about 50 yards from running line	Ground frame Train Tablet	Swanage box		Special services	For working instructions see page 46. *

The Swanage Branch listing of sidings in the Southern Railway working Timetable Appendices (Western Section) for 1934.

SWANAGE.

Beavis' siding.—A 10-ton Portland crane has been provided at a spot situated 30 to 40 feet south of the siding extension, and in connection with the movement of wagons to and from the extension, the special attention of the staff engaged in shunting operations is drawn to Rule 112 (a).

Haysom & Sons' sidings.—These sidings (two), which are laid in partly on private property, consist of a main siding which connects with the Railway Company's siding on the up side at Corfe Castle end of Swanage station, and a spur siding connecting, inside the boundary gate, with the main siding, about 50 yards from the points of the Railway Company's siding.

Wagons containing goods consigned to or from Messrs. Haysom & Sons, and empty wagons for their use, may be placed in or taken from the sidings as required, by arrangement with the Station Master at Swanage, such vehicles to be placed immediately inside the gate, or accepted from that point, as the case may be.

Before wagons are placed in the sidings or taken therefrom, there must be an understanding with the staff of the Stone Company that everything is in readiness for such operations to commence.

Swanage Gas Works siding.—Hand points exist in the reception siding at about 130 yards from the point of connection with the running line, and these points control the entrance to a short lead which divides into two parallel private sidings at a point near the Company's boundary by means of a further set of hand points.

The left-hand siding (going in) is used by the Swanage Gas Company and a turntable is situated thereon at about 100 yards from the siding gate, which turntable connects with a short siding leading to the Gas Company's retort house.

The right-hand siding (going in) is used by the Worth Quarries Company and a loading dock is situated approximately 120 yards from the siding gate.

Wagons for the sidings must be hauled from Swanage with a brake van at the rear in which a competent man must ride.

Upon arrival at the siding the tablet must be obtained from the Driver and the ground frame and siding gate opened. The whole train must then be drawn into the reception siding.

Traffic to and from the Gas Works siding must be exchanged between the boundary gate and the turntable, and traffic to and from the Worth Quarries siding must be exchanged immediately inside the siding gate clear of the fouling point.

The special trip must then return to Swanage by means of a propelling movement with the brake van as the leading vehicle in which the competent man must ride.

During the time traffic is being worked to and from the siding, the Corfe Castle and Swanage section must not be fouled by shunting operations at Swanage.

Extracts from the 1934 Southern Railway Working Timetable (Western Section) Appendices.

Station	Turntables (Length of Rail)		Cranes or runways to lift				Weighbridges			High-way vehicle docks	Water columns	
	Engine	Wagon	Outside		Inside		Truck		Cart		No.	Where situated
	ft. ins.	ft. ins.	T. cwts.	Ht. of lift ft. ins.	cwts.	Ht. of lift ft. ins.	Capacity Tons	Lth. in ft.	Capacity Tons			
SWANAGE BRANCH.												
Corfe Castle ...	—	—	—	—	30	10 3	—	—	—	1	—	—
Swanage ...	50 0	—	5 0B 3 10A	17 3 19 9	30	11 6	20	14	—	1	1	Loco siding

Castle. The Sunday service was meagre — two trains, both in late afternoon; in the morning no wheel turned to distract the inhabitants from their church-going. In the up direction there were also thirteen trains, the 9.50 am, 1.23 pm and 5.27 pm being the London trains. The 7.50 pm was a through train to Bournemouth, running fast from Wareham apart from a stop at Poole. If anyone for London was unwise enough to wait for the last train, the 8.45 pm, they arrived at Waterloo at 4 in the morning. This connection from Wareham was typical of the timetabling mania of the day: on Saturdays it did not call at Holton Heath or Branksome, it only called at Sway on Wednesdays and Saturdays, and it called at New Milton only to set down.

The LSWR was in a monopoly position in Swanage, unlike the situation in Bournemouth or Weymouth; however it was not complete. The steamers from Bournemouth which had served the town before the coming of the railway continued to run, latterly only in summer. Then there was the bus; one guide book in the 'thirties suggested that people from the Midlands and North might prefer to travel by Somerset & Dorset Railway to Poole and take the bus to Swanage. However, this was a tedious journey as it led across the Sandbanks chain-ferry; owing to the steep approach to the ramp, buses had to have their rear ends cut away to avoid grounding on the beach. The SDJR ran up to five through trains from the north to Bournemouth and Poole, and in some years there was a through carriage for Swanage; with a change at Poole the journey by this route from Birmingham to Swanage then took well under six hours and was a good deal more convenient than travelling through London.

In one way however the branch was unfortunate. When the LSWR began its halt-building programme in 1903 one might have expected one or two in Purbeck. But to be truthful, there was nothing to "halt" at. Owing partly to geography but mainly to the conservative attitude of landowners, notably the Bankes family, no holiday villages ever developed in the peninsular.

The nineteen-thirties were perhaps the best years for the Swanage Branch. Holiday traffic was increasing fast, and private cars had not yet made great inroads. In addition, the area had always been a favoured one for Territorial Army camps — so much heath and down to put tents on — and with the increasing likelihood of War, these were also increasing. The limit for troop trains seems to have been ten coaches plus extras (horse-boxes, field-kitchens, sometimes guns) and these trains took some starting up the bank out of Swanage. Western Section engines in those days carried their power classification letter on the front-end of the running plate. A small Drummond 4–4–0 classified 'F' and an Adams 4–4–0 classified 'I' double-heading a full train seemed to get away fairly well; another train headed by two 'M7' 0–4–4Ts (classified 'K') did, as expected, have a hard struggle. It was particularly difficult if the main train had to set back after leaving the platform to pick up baggage vehicles from the goods yard, as it gave no space for a run at the bank. The type of engine which could be stabled at Swanage was of course limited by the length of the turntable. From photographs, it would appear that some time after 1938 an entry was made into the engine shed from the Wareham end, thus enabling an engine too long or heavy for the turntable to use the shed facilities.

The War did not greatly affect the branch; for several reasons this part of the coast was not a very attractive invasion choice, however the 14th Super Heavy Battery, 5th Corps, set up rail-mounted long-range guns on short spur lines, one 300 yds on the Swanage side of Furzebrook, another ½ m the other side (about October 1940) and they stayed for some ten months. The SR provided 'K10' class 4–4–0 No. 393 to service them.

Afterwards, petrol rationing brought a short term benefit to the passenger traffic, but although visitors to Swanage itself were to increase by leaps and bounds, the railway began a gradual decline, particularly in terms of through trains.

In 1951 the newly created train "The Royal Wessex" appeared, a short-lived return to the old days of named trains on the former Southern. This was a triple-portion train, with five coaches for Weymouth in front, then two coaches for Swanage, and six coaches for Bournemouth at the rear. It left Waterloo at 4.35 pm and the two coaches arrived in Swanage at 7.52, having been detached from the Weymouth train at Wareham and attached at the rear of the branch push-and-pull; normally the branch train drew forward from the south bay and backed on to the through coaches in the down through platform. The return working was 7.38 am from Swanage.

In 1955 the service from London to Swanage was reasonable and comprised:

Waterloo	1.14 am	Swanage	6.48	via Salisbury, 3rd class
,,	2.40	,,	8.40	3rd class, ½ hr wait at Wareham
,,	5.40	,,	10.09	
,,	8.30	,,	12.19	
,,	10.30	,,	1.54	Through carriages
,,	11.30	,,	3.59	Not Sats
,,	12.30 pm	,,	3.59	"Bournemouth Belle" to Central
,,	12.35	,,	3.59	Sats only
,,	1.30	,,	5.21	
,,	3.20	,,	6.49	
,,	4.35	,,	7.52	"Royal Wessex"; through carriages
,,	6.30	,,	9.42	Through carriages

There were four other arrivals at Swanage: 8.53 am from Wareham, 12.38 pm with connection from Southampton, 2.38 pm from Wareham only, and 6 pm from Bournemouth.

In the early sixties there was one through coach for Swanage in the winter service, normally formed as the fifth coach of a thirteen-coach train from Waterloo. At Bournemouth Central the rear eight coaches would be removed, and at Wareham the fifth coach would be left in the platform, to be picked up by the push-pull train for Swanage.

The "Pines Express" was shown in Bradshaw as connecting for Swanage. Times were not very exciting: 11 hours from Bradford, 8 hours 36 mins. from Manchester, 6 hours 11 mins. from Birmingham; but for holidaymakers with large families it was still worth while to avoid the change in London,

The branch train a few weeks after the opening hauled by Beattie 2–4–0T No. 209.
S.R. Magazine

Swanage station on 6th June, 1930, with 'T1' 0–4–4T No. E67 in charge.
H.C. Casserley

The locomotive shed on 7th August, 1932; No. E70 prepares for the Sunday afternoon-only service. Author

A branch train entering Swanage *c*.1906. *Lens of Sutton*

Down train entering Corfe Castle station about 1905 hauled by LSWR 4–4–0 No. 477.
Author's collection

Corfe Castle station looking north, about 1930. *Lens of Sutton*

An up goods and empty stock train near Eldon Siding in 1938; note Fayle's engine on left shunting wagons at a weathering bed. *Author*

Driver Whipple (*left*) and his fireman pose on the branch engine, No. E62, at Swanage in 1932. *Author*

Up troop train passing through the Corfe Gap 11th August, 1934; 'K10' class 4–4–0 No. E387 is piloting an Adams "seven-footer" *Author*

An up London train near Furzebrook with 'X6' class 4–4–0 No. E658 piloting an 'M7' 0–4–4T on 10th August, 1932. *Author*

Two 'M7' class engines head a long troop train (the 4th Gloucesters) out of Swanage on 15th August, 1932. *Author*

The branch train leaves Swanage on 15th August, 1932 hauled by 'T1' class 0–4–4T No. E73. *Author*

On 24th November, 1976 a clay train sets off "wrong line" for Wareham at Worgret Junction, after the lifting of the crossover. *Author*

A train of clay wagons leaves Furzebrook on 12th July, 1978, with the newly-constructed oil and gas terminal on the left. *Author*

Swanage seen on 1st September, 1968; the bay line and engine loop are already removed; DMU No. 1114 waits for departure time. *D. Gould*

Wareham station with Locomotive No. 133 awaiting to depart to Swanage. *c.*1906. *Author*

the change at Poole being a good deal easier. These timings had been made worse as against pre-War ones by arranging a 40 minute wait at Poole and another of ten minutes at Wareham.

After "dieselisation" in 1966 the service was altered to allow only one DMU to be used; the weekday roster of this train is on the next page; one may wonder why a diesel fuel depot was not set up at Swanage enabling the train to be stabled there.

Meanwhile the working of the Bournemouth line had been revolutionised by a system of express push-pull working, using a combination of motored 4-REP electric sets, 3 and 4 car non-motored sets, and diesel and electro-diesel locomotives. Any combination might be seen; indeed a train with a 4-REP set leading, followed by 4 or 8 unmotored coaches, with a diesel locomotive behind, was quite common. The Swanage branch was catered for by those unmotored 4-TC sets; in May 1969 for instance a set was attached on weekdays to the 15.30 from Waterloo; this reached Swanage at 18.10, had four minutes lay-over, and departed at 18.14 for Bournemouth. On Summer Saturdays a set left Waterloo with the 07.57, arrived at Swanage at 10.32, and left at 10.37 arriving at Waterloo at 13.21 attached to a Bournemouth express. The timing of 2 hours 35 minutes compares with the 3 hours 32 minutes of the 9.18 am through portion of the early 'thirties.

The branch was not on the Beeching list of lines destined for the axe, but it seems second thoughts prevailed, for in 1967 it was proposed by British Rail that it should be closed from 9th September, 1968 to passenger service, and should become a freight line serving the clay sidings only. Local opposition to the closure was successful in giving the branch a short further lease of life, but in a climate of widespread closures in other parts of the Region the end could not be long delayed. It was scarcely likely to encourage holiday traffic when the Southern Region's timetables for the period May 1968 to May 1969 proclaimed that the service might be withdrawn during the currency of the timetables! The through carriages ceased; the last runs were the 07.39 up from Swanage and 15.30 down from Waterloo on 3rd October, 1969. However, local authorities were dissatisfied with the alternative bus services proposed, and succeeded in delaying the final closure until the beginning of 1972; it was in fact on 3rd January of that year that the last passenger train left Swanage, from a station little more than a ghost, with the engine release road and sidings long gone, the turntable overgrown, and only the fine stone station (mutilated at some stage by the replacement of the LSWR canopy with a convex SR design) to remind one of the great days when only the best was good enough.

However, there were many people who were not prepared to confine their activities to mourning. The preservationists were soon in full cry, with the blessing of Sir John Betjeman, for whom Swanage must surely have represented the epitome of Victorian family holidaymaking which he was recalling so forcefully in print and on television.

The Swanage Railway Society obtained a good press, a little patronising perhaps locally, and they got down to cases. British Rail were asked if they would work trains forward from Worgret; if they would not, then a parallel line from Worgret to Wareham south bay would have to be built. Additional halts would be put in, at Worgret, Blue Pool, Harman's Cross, and Herston. At least eight trains a day would be run, and a special school train would take 400 children from Swanage to the new "Comprehensive" at Worgret. A coal terminal would be put in at Swanage, and a terminal for the stone traffic at Harman's Cross.

Resulting from a misunderstanding between the SR estates department and the Swanage Railway Society, contractors began lifting some track in the summer of 1972. Telegrams were sent to various authorities to halt the work, and a petition signed by 2500 residents was sent to the Minister. A "Fighting Fund" was also set up. A further misunderstanding caused the destruction of the water tower at Wareham, local officials having concluded that if the Swanage Railway did come about, it would use diesels. By the end of 1972, the railway authorities had informed Dorset County Council and the Society that they had only until February to come to a final decision. At the same time, it became known that a property company had made a bid for the site of Swanage station, without which the Swanage Railway could hardly succeed, though contingency plans were made for setting up a terminus beyond Victoria Bridge.

Early in 1973 it was revealed that the local authorities had a plan for building a Corfe by-pass along the route of the railway. The preservationists countered with a suggestion to build a tunnel for the by-pass through the down at the east side of the railway leaving the railway cutting untouched.

At the same time work began on restoring the locomotive *Swanage*, a Bulleid Pacific which after withdrawal had been residing in the scrap-yards at Barry Docks in S. Wales. This engine, No. 34105 in BR days, was not the only one of its class to survive, as another had been at the Liss steam centre and was at this time at the Bluebell Railway depot at Sheffield Park, but the fact that the engine was named *Swanage* provided a rationale for tying it up with the general preservation drive.

However, by the Spring of 1973 the branch track had been lifted, except for the stretch between Worgret Junction and a point 300 yds south of Furzebrook sidings. The further history of the "new" Swanage Railway is detailed in *Chapter Five*.

The DMU diagram for 1966.

Arrive		**Depart**
—	Middle Siding No. 1	6.18
6.20	Bournemouth	6.30
7.38½	Weymouth	7.45
8.17	Wareham	8.23½
8.43	Swanage	8.54
9.14	Wareham	9.20
9.39½	Swanage	9.48
10.08	Wareham	10.48
11.07½	Swanage	11.45
12.05	Wareham	12.41
13.00½	Swanage	13.48
14.08	Wareham	14.41
15.00½	Swanage	15.10
15.30	Wareham	15.35
15.45½	Swanage	16.16
16.36	Wareham	16.44
17.03½	Swanage	17.32
18.20½	Bournemouth	18.25
19.12½	Swanage	19.48
20.08	Wareham	20.43
21.02½	Swanage	21.15
22.02	Bournemouth	
	Wednesdays only	22.04
22.47	Eastleigh	22.50
23.06	Eastleigh Diesel Depot	
	Wednesdays Excepted	
22.02	Bournemouth	22.30
22.41	Carriage Sidings WFX	23.30
23.40	Middle Siding No. 1	

Chapter Two

Locomotives and Rolling Stock

It would seem that at first the branch was worked by the well-known "Beattie Well Tanks", as a photograph exists of No. 209 (of that class) on the train soon after the opening. Even then, however, these engines, which had been built by Beyer Peacock some 20 years earlier, were thought not to be up to this type of work, and it is probable that they were replaced fairly soon by one of the classes 4–4–2T. There were three classes of this type, the "Ironclads" which were rebuilt 4–4–0Ts, the "steam rollers" with solid bogie wheels, and the class later referred to as the "0522", comprising engines built by a great variety of makers: Beyer Peacock, Stephensons, Dübs, and Neilson, around 1882–5.

However although the "branch engine" itself would have been first a 4–4–2T, then a 'T1' 0–4–4T, later a 'M7' 0–4–4T (from about 1932) and finally a BR 2–6–2T or 2–6–4T, the through trains of all kinds which found their way to Swanage were hauled by a wide range of classes. At first the 'A12' 0–4–2 was probably a frequent visitor; this class was begun in 1893. The graceful outside cylinder 'X2' and 'T3' 4–4–0 begun a year earlier certainly appeared, and also the similar 'T6' and 'X6' built three years later. After the turn of the century more powerful engines with smaller wheels were preferred: the 'L11' and 'K10' 4–4–0 by Drummond. In the 'thirties the 'T9' 4–4–0 sometimes appeared. This numerous class had been started in 1899 for express work, and when Drummond turned his attention to the 4–6–0, these engines descended to lighter work. The smaller 4–4–0s remained more common on the Swanage Branch, however, and because of the severe gradients they were very often worked in tandem, or assisted by a tank engine.

Goods traffic was sometimes worked by the '700' class 0–6–0 designed by Drummond and built from 1907 on, though the 'L11' and 'K10' 4–4–0s with only 5 ft 7 in. wheels were equally capable — not that there was much hard work to be done latterly, for the loaded clay wagons would be attached to the train only after most of the climbing had been carried out. No doubt in the early days a fair amount of stone was carried from Swanage, and to lift that up to the "Corfe Gap" would call for some collar work. In 1938 the new 'Q' class 0–6–0 appeared on the line, but usually in passenger service.

In the early sixties the 'M7' class began to be joined by Ivatt and BR 2–6–2T and 2–6–4T engines, with occasional visits from BR 2–6–0s, and the branch showed its freedom from cramping restrictions by accepting also the occasional Bulleid "West Country" Pacific.

Steam had had a good run, but on 5th September, 1966 the arrival in service of "Hampshire" 3-coach DMU No. 1104 heralded its end. The last fling was probably the "Dorset Coast Express", a special which ran down and up the branch on 7th May, 1967 on its way from London to Weymouth, hauled by Bulleid Pacific No. 34023, assisted by 2–6–4T No. 80011. The through trains, now restricted in number, were worked by various classes of diesel locomotive, which also worked the clay traffic, for the short period before passenger closure.

For the first few years of the branch, four-wheeled coaches were used, but by about 1895 bogie coaches would have been appearing on through workings, and soon after 1900 the standard two-coach non-corridor bogie sets would be available, though six-wheeled stock probably also appeared until the 'twenties. The 2–8 sets on the branch in the early 'thirties normally had a brake compartment at each end, and being fairly short coaches anyway, the seating capacity was small. Whenever there was likely to be extra traffic, loose coaches would be added; by the 'thirties these were often corridor coaches. The through workings at that time were a mixture of 2-, 3-, 4-coach corridor sets of the later LSWR or the first Maunsell type. A number of trains worked through in the summer from the LMS via the Somerset & Dorset, and the author noted in 1932 on a Girl Guide's Special, an LMS low-roofed twelve-wheeler — not the kind of thing to be found at the end of many branch lines! With the electrification of lines around Aldershot, more of the ex-LSW 2–8 push-pull sets became available, and these were often on the branch workings, with suitably-fitted 'M7' 0–4–4Ts, until about 1964, when the 2–6–2Ts which were arriving were not fitted for working them. If as often happened the engine was at the Wareham end, the train after connecting with the through London coaches would comprise the push-pull set leading, then the engine, then the through coaches. In the last days of steam various types of coach appeared, including one of SE&CR origin.

The last batch of 'M7' class engines noted as rostered for Swanage duties, which were alternated with the Brockenhurst branch workings, were Nos. 30031, 30108, 30110, 30112, 30328. The push-and-pull trains which they worked were now made up from Maunsell corridor coaches in the main.

The final type of rolling stock on the branch, apart from the DMU, was the 4-coach unmotored set designed for working in unison with the 4-REP electric sets which worked from Waterloo to Bournemouth. However, some of the most modern locomotive hauled stock did make its appearance at Swanage, when the "Royal Wessex" train from Waterloo to Weymouth was instituted in 1951, with a Swanage portion, as mentioned previously. How sad that a station which could boast of coaches bearing these proud roof-boards standing at its platform should so soon have declined to a state that, when one went "to see the train", there was never anything there but a moaning and juddering DMU, and in only a few years more — no train at all.

A Pullman "Holiday Coach" (No. P.45 former *Ruby*) was placed at the end of Corfe Castle sidings in the 'sixties.

It should be noted that a special goods brake was kept at Wareham for the clay traffic: Maunsell 25-ton No. 55687, lettered "To work between Wareham and Furzebrook. Not Common User". This was still in evidence in 1973.

WAREHAM and SWANAGE. – London and South Western.

Miles	Down. Week Days.	mrn	mrn	mrn	mrn	mrn	aft	aft	aft	aft	aft	Thurs. aft							Sundays. aft	aft		
	Waterloo Station, 126London ....dep.	....	5 50	6 35	1015	1040	1230	2 0	2 10	4 10	4 50	6 55	....	....	....	....	....	....	1210	6 40	....	....
—	Wareham .... dep.	7 53	10 0	1148	1 25	3 20	3 49	4 53	5 44	6 50	8 15	1150	....	....	....	....	....	....	3 47	1011	....	....
6	Corfe Castle	8 6	1013	12 1	1 38	3 33	4 2	5 6	5 57	7 3	8 28	12 3	....	....	....	....	....	....	4 0	1024	....	....
11	Swanage.. .... arr.	8 20	1024	1212	1 49	3 44	4 13	5 17	6 12	7 14	8 39	1214	....	....	....	....	....	....	4 11	1035	....	....

Mls	Up.	mrn	mrn	mrn	mrn	aft	Thurs. b	aft	aft	aft	aft	aft	Thurs. aft	Sundays. aft	aft		
—	Swanage..........dep.	7 28	7 58	8 57	1045	1243	2 0	2 57	4 20	4 53	5 50	7 35	9 45	2 45	5 42	....	....
5	Corfe Castle	7 42	8 9	9 9	1057	1255	2 11	3 9	4 32	5 8	6 5	7 47	9 57	2 57	5 54	....	....
11	Wareham126,131arr.	7 54	8 20	9 21	11 9	1 7	....	3 21	4 44	5 20	6 17	7 59	10 9	3 9	6 6	....	....
132	131London (Wat.)arr.	....	11 8	1246	2 45	4 9	....	6 45	7 50	....	10 5	1159	3 35	8 6	9 29	....	....

b Runs to Bournemouth West, see page 132.

The branch passenger timetable for April 1910.

WAREHAM and SWANAGE. – London and South Western.

Miles	Week Days.	mrn	mrn	mrn	mrn	mrn	mrn	aft	aft	aft	aft	aft	aft	Thurs. aft					Sundays. non	aft		
	Waterloo Station, 136London ....dep.	....	....	5 50	6 35	1015	1025	1230	1240	2 0	2 10	4 10	4 50	6 55	....	....	....	....	12 0	....	....	....
—	Wareham .... dep.	8 0	a	10 0	1148	1 25	3 20	3 49	....	4 39	5 44	6 44	8 15	11 50	....	....	....	....	3 48	6 25	....	....
6	Corfe Castle	8 13	8 43	1013	12 1	1 38	3 33	4 2	....	4 52	5 57	6 57	8 28	12 3	....	....	....	....	4 1	6 38	....	....
11	Swanage.. .... arr.	8 29	8 54	1027	1212	1 49	3 44	4 13	3 50	5 3	6 12	7 8	8 39	12 14	....	....	....	....	4 12	6 49	....	....

Mls	Up.	mrn	mrn	mrn	mrn	mrn	aft	Thurs. aft	aft	aft	aft	aft	aft	Thurs. aft				Sundays. aft	aft		
—	Swanage..........dep.	7 29	8 7	8 58	10 5	10 45	12 45	2 0	2 56	4 17	4 40	5 50	7 35	9 45	....	....	....	2 45	5 43	....	....
5	Corfe Castle	7 43	8 18	9 12	1016	10 57	12 57	211	3 8	4 29	4 54	6 5	7 47	9 57	....	....	....	2 57	5 55	....	....
11	Wareham136,141arr.	7 55	8 29	9 24	1028	11 9	1 9	b	3 20	4 41	5 6	6 17	7 59	10 9	....	....	....	3 9	6 7	....	....
132	141London (Wat.)arr.	....	11 8	1245	2 3	2 45	4 0	...	6 45	7 44	....	10 5	12 4	b	....	....	....	8 6	9 39	....	....

a Runs from Bournemouth West, see page 136. **b** Runs to Bournemouth West, see pages 142 & 143.

Passenger timetable for October 1911.

WAREHAM and SWANAGE.—London and South Western.

Miles	Down. Week Days.	mrn	mrn	a	mrn	a	mrn	a	mrn		aft	aft	a	aft	aft	aft	aft		Sundays. mrn	mrn	aft	
	Waterloo Station. 134London ....dep	....	....	5 40	5 40	8 0	8 30	1030	9 30	....	1230	....	2 30	1 30	4 30	a	6 30	....	....	11 0	2 0	...
—	Wareham ....dep	8 0	9 10	9 46	1020	1143	12 4	1 21	2 4	....	3 24	4 35	5 24	5 46	7 29	8 10	9 23	....	1122	3 12	6 12	...
6	Corfe Castle..................	8 13	9 23	9 59	1033	1156	1217	1 37	2 17	....	3 37	4 48	5 37	5 58	7 42	8 23	9 36	....	1135	3 25	6 25	...
11	Swanage ....arr.	8 26	9 37	1014	1044	12 7	1229	1 50	2 28	....	3 50	5 0	5 50	6 7	7 55	8 34	9 47	....	1146	3 36	6 37	...

Miles	Up. Week Days.	mrn	mrn	a	mrn	mrn	mrn	a	a	aft	aft	a	aft	aft		aft	a	aft	Sundays. mrn	aft	aft	
—	Swanage ....dep	7 0	8 0	8 45	9 15	9 50	1055	11 20	1240	1 25	2 15	3 2[illegible]	4 35	5 25	....	6 23	7 30	8 40	1015	2 10	5 15	...
5	Corfe Castle	7 11	8 11	8 55	9 26	10 1	11 6	11 31	1251	1 36	2 46	3 [illegible]	4 46	5 36	....	6 36	7 41	8 51	1026	2 21	5 26	...
11	Wareham 134, 137 ...arr.	7 24	8 27	9 10	9 41	1015	1119	11 45	1 4	1 50	3 0	3 50	5 6	5 50	....	6 50	7 55	9 4	1039	2 34	5 43	...
132	137London (Waterloo arr.	11 0	1222	....	1250	2 20	4 [illegible]0	2 50	...	4 50	...	6 50	....	8 50	...	1128	....	3 55	2 16	6 22	9 26	...

a Commences on the 14th instant.

Passenger timetable for July 1922.

SWANAGE (Dorset). 131 miles. From Waterloo. Fares, 27/6a, 16/6c; Return, 55/0a, 33/0c. E. Clos. Day: Thurs. Pop. 4,690.

W'loo	Swan.	Swan.	W'loo.
AM5 40	1045	AM7 35r	1059
8 30r	1229	9 10r	1249
9 30r	2 29	9 43r	2 20
PM1230r	3 56	1053kr	2 54
1 30r	6 30	PM1 20r	4 50
2 30kr	6 30	2 45r	8 24
4 30r	7 55	5 31r	8 55
6 30r	9 49	6 35	1132
—	—	8 40	3 58

SUNDAYS.

PM2 0r	6 40	PM5 15r	9 42
—	—	—	—
—	—	—	—
—	—	—	—

k—Mons., Fris. & Sats. only
n—Saturdays excepted:
r—Restaurant Car.
s—Saturdays only.

Passenger timetable for April 1927

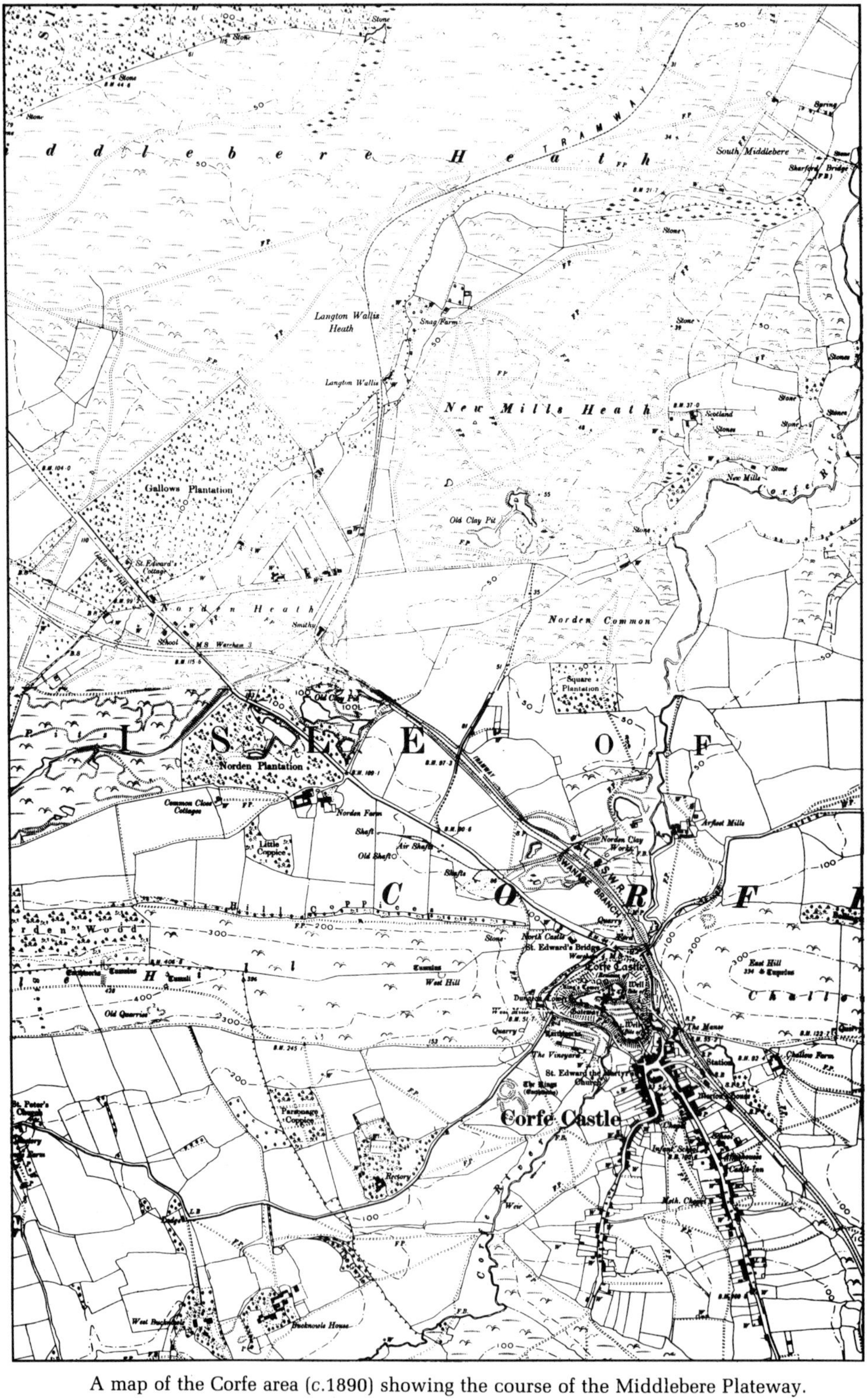

A map of the Corfe area (c.1890) showing the course of the Middlebere Plateway.

Chapter Three

The Clay Tramways

Ball clay suitable for fine china is rare in England, and the deposits in Purbeck are some of the finest, as well as being suited in most cases for open-cast working. Although some small-scale digging went on earlier, it was in 1760 when the Pike Brothers purchased the Furzebrook deposits that commercial work began. In 1795 Benjamin Fayle acquired the Norden area somewhat south of Furzebrook, and these two names were to dominate the trade for a century and a half. There was of course no call for clay in that area, which meant transport to places such as the Potteries in Staffordshire. Wareham was a port of sorts and carting to there was possible, but Fayle in 1806 took advantage of the developing art of plateway construction, and had one laid for him to a creek off Poole Harbour. The Pikes did the same thing about 1840, but to a wharf further west and on the main Wareham Channel. Clay is particularly in need of the railway or plateway facility, as it has to be handled twice. First it is taken from the pits to weathering beds, where it stays for a couple of months, being turned, and then when ready is loaded again for transport to ships or main-line railways.

There were four main rail systems in the clayfields: the Middlebere Plateway; Fayle's Tramway, which was a modernised version and also took account of the arrival of the Swanage Railway; the Newton Tramway, also owned by Fayle, which exploited two enormous open-cast workings near the now-extinct village of Newton, and was later joined up with Fayles Tramway; and Pikes Tramway which, although in origin a simple straight line from the present "Blue Pool" to Ridge, ended up with a very complex system covering all parts of the Creech Beds.

THE MIDDLEBERE PLATEWAY

This was built by Benjamin Fayle from his pits at Norden near Corfe Castle to a wharf on Middlebere Creek, a channel running south from the Wytch Channel at Poole Harbour, which must then have been navigable by ships at some states of the tide. The plateway pursued a winding course to provide a level enough track for horse-haulage, though some shallow cutting was required in the middle part of Middlebere Heath, and tunnels under the Wareham Road as described later.

Although plateways and railways were becoming known in the north, they were rare in the south; when the Swanage peninsular was mapped in 1811 by Lt-Col Madge, the only sign of industrial activity he noted was the "Iron Rail Way" to Middlebere; he spotted two passing loops, and it would be interesting to know the thoughts of an old soldier of that time about this new-fangled thing.

The Middlbere line is first described in *A General View of the Agriculture of Dorset* by William Stevenson, published in 1812. He states that it was built in 1806 and was 3½ miles long; the tramplates were three feet long, the vertical and horizontal sides both measuring three inches. The stone sleepers weighed 60–70 lb each, and the plates were fastened with one spike in

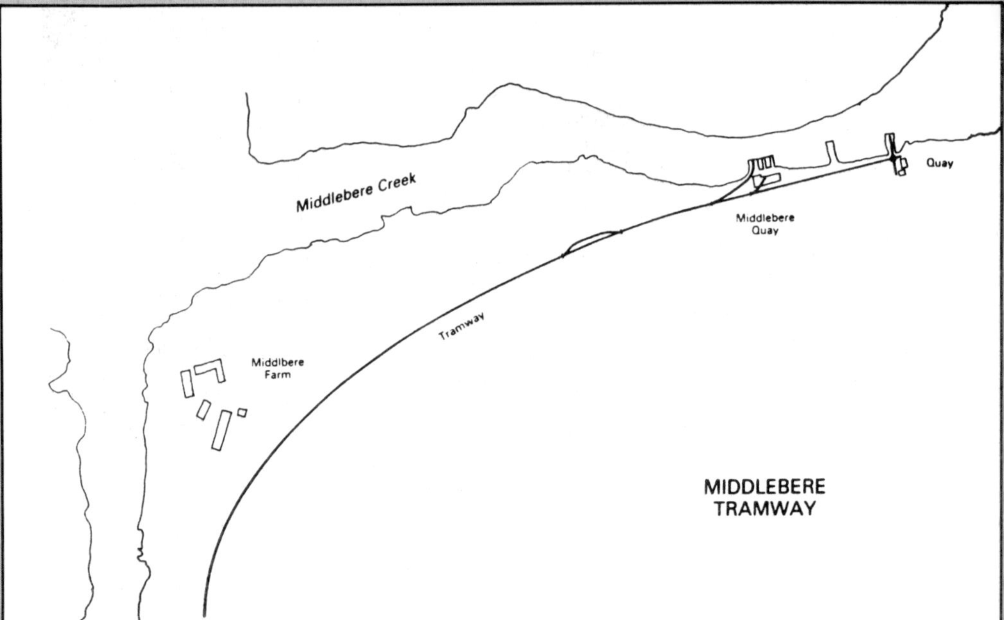

Northern end of the plateway of the Middlebere Tramway.

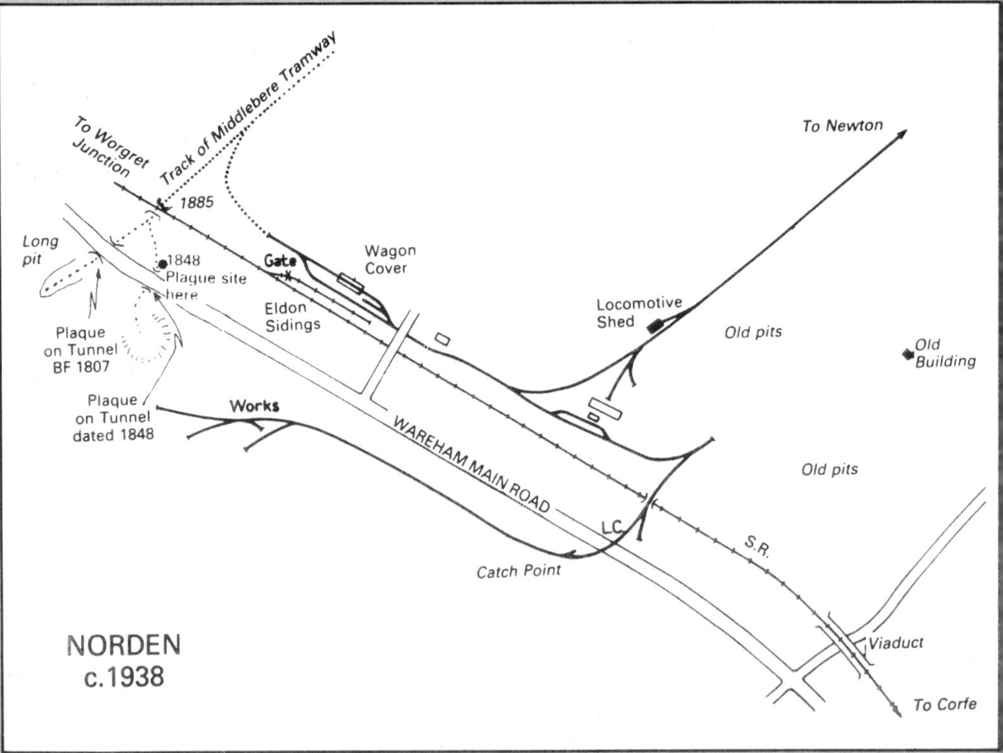

A 'Q' class 0-6-0 propels a down goods across the viaduct spanning the Studland "New Road". The area on the left beyond was worked by Fayles Tramway at the turn of the century, 3rd July, 1955. *J.R. Bonser*

Weymouth (left) and Swanage trains in Wareham station in June 1964; 2–6–4T No. 80147 is on the 3.37 pm to Swanage. *J. Scrace*

2–6–0 No. 76013 climbing away from Swanage with a goods train in June 1964. *J. Scrace*

The branch train in Swanage station in June 1964; 2–6–2T No. 41238 simmers with a Maunsell 2–8 set and extra coach. *J. Scrace*

The outside view of Corfe station from the approach road, 3rd July, 1955. *J.R. Bonser*

DMU No. 1128 on the 2.41 pm Wareham–Swanage train at Corfe Castle in August 1969. *J. Scrace*

TC unit No. 427 hauled by D6538 forms the 6.14 pm Swanage–Bournemouth train on 7th August, 1969, here seen entering Corfe Castle. *J. Scrace*

Activity at Swanage on 21st October, 1979; the ex-GWR hand-crane is moving rail stocks, propelled by the diesel locomotive *May*. *Author*

A train hauled by 0–6–0ST No. 21 at Herston station in July 1985. *A.K. Pfrangley*

A general view of Swanage station in July 1985; the steam engine in the platform is ex–MR 0–6–0T No. 41708 on loan from Butterley. *A.K. Pfrangley*

an oak plug. The gradient is given as 1 in 150 to 1 in 180, and the whole work cost £2000 per mile. The wagons carried two tons of clay each and three horses drew five wagons at a cost of 6 *d*. per ton.

An isolated mention in Farey's *General View of Agriculture in Derbyshire* (1817) notes that "Collinge's patent axletree has been applied to the trams on the Rail-ways to the famous pipe-clay pits near Corfe Castle". This axle, patented in 1811, was largely used for road vehicles and no other application to railway vehicles is known. It implies that the wheels revolved on fixed axles, which was commonly the case with plateway vehicles, some examples of which from other plateways have been preserved.

Stevenson stated that the clay was being shipped from Middlebere in sea-going craft to places such as Liverpool, and no doubt the channel was deeper in those days; by the end of the century only craft of some 30 tons could use the creek.

The plateway originated at some clay pits 1 mile north-west of Corfe Castle on either side of the main Corfe–Wareham road, and shortly after opening it was necessary to tunnel under this road. There are in fact two tunnels; the northern one has a plaque on the east face reading "BF 1807"; the southern one has on its west face "Dated in 1848". As this tunnel appears on earlier tithe maps this must be a rebuilding date. Both tunnels are very low, and the horses had to have special harnesses which did not project upwards at all. The remains of both tracks now appear to pass through one skew tunnel under the LSWR, but as the southern track has a very serpentine course it probably originally made junction the other side of the LSWR at the smithy and washing beds near New Line Farm, but was slewed to allow the LSWR to build only one tunnel under itself.

At the Middlebere end there was a siding and turntable on the wharf, and a siding to a slipway. There was also a flag-pole; this could have been to signal to a ship in the main channel to come in, or it could have been to let someone inland know that a ship was putting in.

By the mid-eighties it seems that the pits south of the Wareham Road were worked out, and a line was run from the washing beds parallel with the LSWR to some new pits about a quarter of a mile south, the line ending in an incline as the new pits were down at the level of the Corfe River. From the top of this incline another line ran back across the LSWR to pits between it and the main road, and later across the road to the area of pits and shafts which still exists at Norden. Whether this trackage was plateway or edge rail is in doubt. The Ordnance Survey of 1885 shows a physical connection at the washing beds; this could be an error, in which case the Norden section could have been edge rail and the Middlebere section still plateway. However, if one accepts this connection, it follows either that the new tracks were plateway, or that the Middlebere line had been relaid in edge rail. It must be said that although the route to Middlebere yields many one-hole stone block chairs, it gives no hint of sleepers or two-hole blocks, as edge rail would have required. The standard work on plateways by Bertram Baxter gives no help as it erroneously states that the Middlebere Plateway was closed when Pike's line was opened.

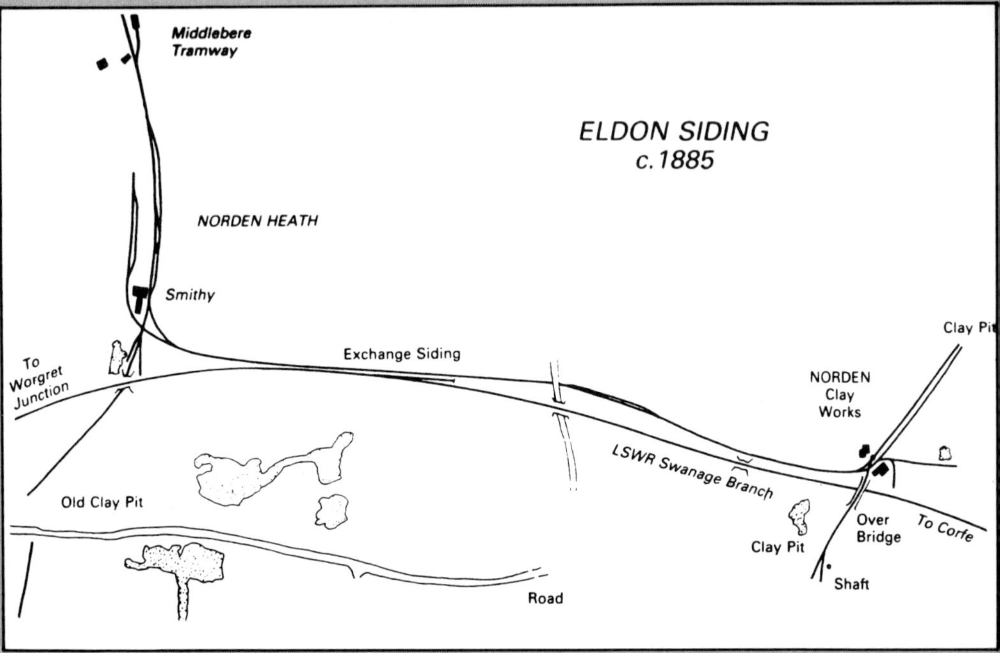
ELDON SIDING
c.1885
Middlebere
Tramway
NORDEN HEATH
Smithy
To
Worgret
Junction
Exchange Siding
Clay Pit
NORDEN
Clay
Works
LSWR Swanage Branch
Old Clay Pit
Clay Pit
Over
Bridge
To Corfe
Shaft
Road

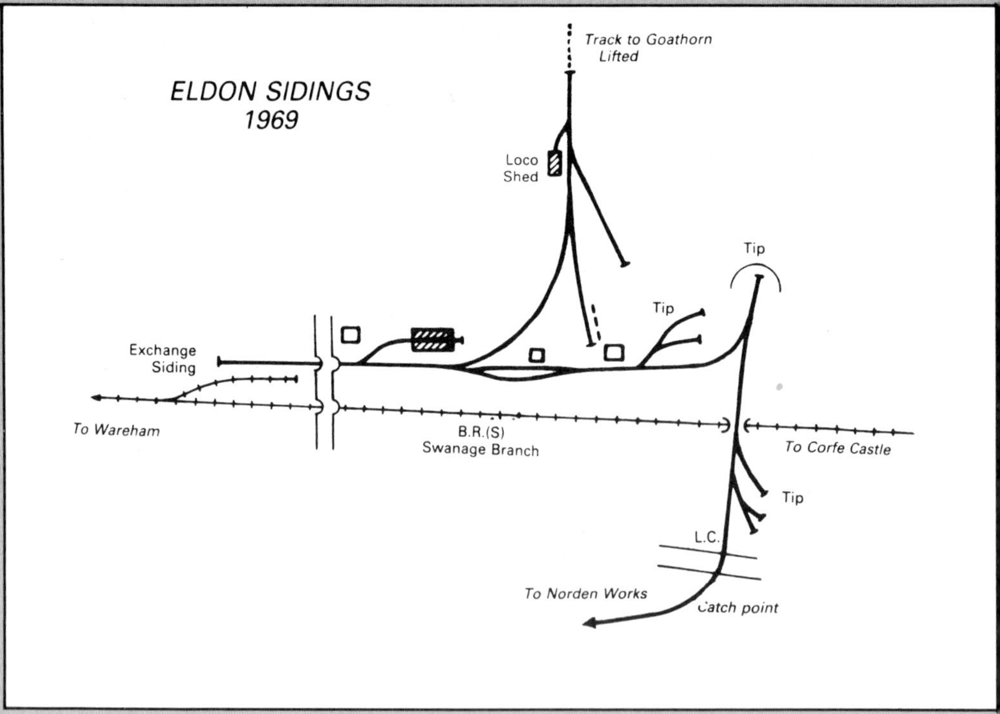
ELDON SIDINGS
1969
Track to Goathorn
Lifted
Loco
Shed
Tip
Tip
Exchange
Siding
To Wareham
B.R.(S)
Swanage Branch
To Corfe Castle
Tip
L.C.
To Norden Works
Catch point

There is a third tunnel on the system, under the Studland Road, giving access from a pit on East Hill to the rest of the complex. This does not appear on the 1886 OS map; the connection's exact course cannot be determined owing to dumping and digging in the area.

THE NEWTON TRAMWAY

This short line, not much over a mile long, was opened by Fayle about 1868 and ran from two very long open-cast pits on Newton Heath through the attractive Goathorn Plantation on its oddly shaped peninsular to a three line layout at Goathorn Depot, which narrowed to two tracks as it approached the wooden pier, these converging at the end at a metal loading shute. The gauge of the line was in later years 3ft 9in. but is said to have originally been 3ft 6in., being re-gauged to suit the Lewin engine, presumably after the connecting line to Norden was laid, in 1904. It was probably horse-worked, though the Lewin engine may have worked there. On a visit in 1938 a building was noted which might have been an engine shed. Working ceased about 1936, and the pits, one some half-mile long, are hard to find in the forestry area; the Newton works was totally destroyed during the War.

FAYLE'S TRAMWAY

This is the title given to the completed 5¾-mile-long system stretching from the Norden pits to Goathorn Pier. By the turn of the century the south-western end of the Middlebere Tramway had become disused; as stated previously, a branch from the smithy running south-east towards Corfe was in operation, and another branch from this ran over the LSWR Swanage branch to pits again west of the Wareham Road, but south of the original ones. Some time around 1905, not necessarily both at the same time, two changes were made: an exchange siding was set up between the clay line and a single-line siding from the LSWR (Eldon Siding), the clay line running on a stage so that clay could be loaded easily into the main line wagons. From a little further north-east a line was built over the sandy heath to meet the Newton Tramway at a point just west of those workings. Thus Norden now had an outlet to Goathorn Pier and one to the main-line railway, two ways in which the clay could be shipped away.

By 1905 Fayle's Tramway was using normal railway track, to a gauge of 3ft 9in., and had two steam locomotives. One was a very small 0–4–0T built by Stephen Lewin of Poole about 1870, said to have originally been named *Corfe* and later *Tiny*, though it bore no name by 1930. The other was a larger Manning Wardle 0–4–0ST (Works No. 1552 built 1902) purchased in 1909 from the Northern Outfall sewerage line of the LCC at Barking, and converted from 3ft 6in. gauge. This must have been purchased for the long haul to Goathorn, as it was too high to pass under the road bridge between Norden depot and the exchange sidings. (In 1938 the track was lowered under the bridge to enable it to do so). This engine retained its name *Thames* on cast plates on the saddle tanks and number 48 on cab-side. This was the engine used to run the line's "passenger service" from 1920 to 1936 after

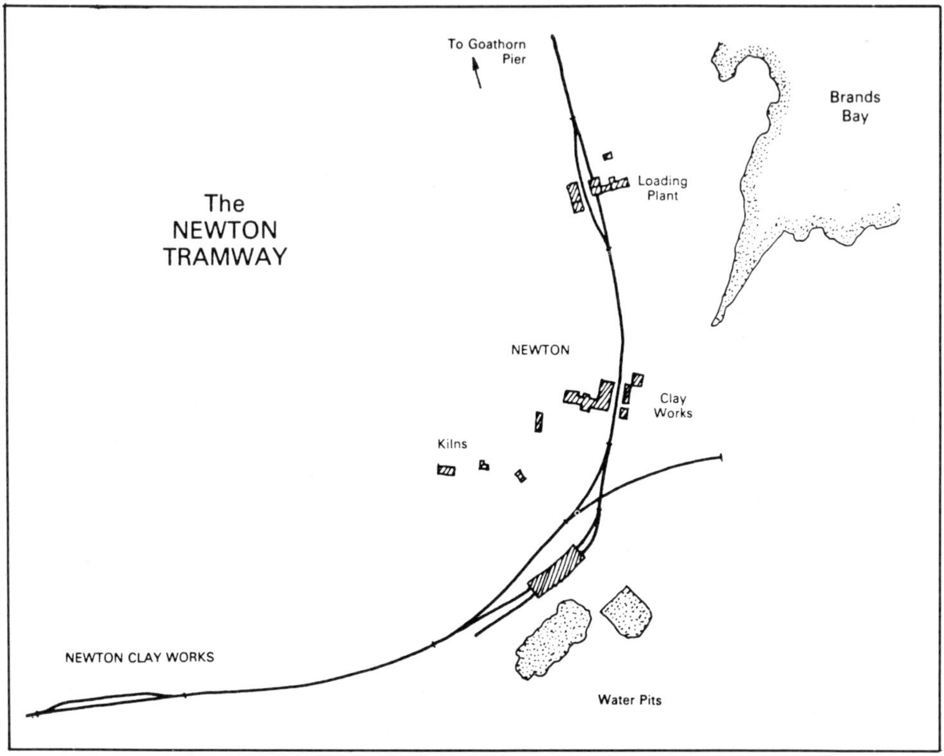

Two long clay pits extended south-west from the depot. These were probably served only by light rails suitable for horse working; c.1890.

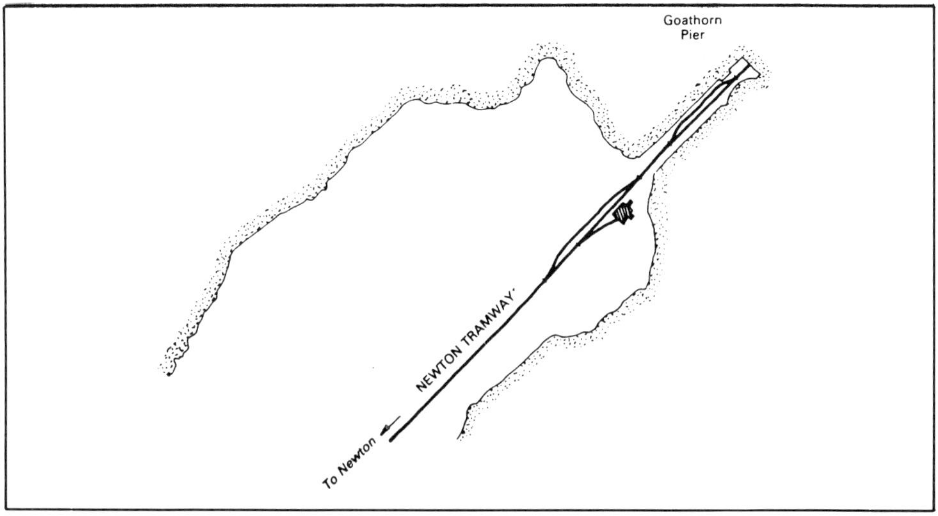

Newton school was converted to a chapel. *Thames* and a converted clay wagon with a corrugated-iron roof conveyed schoolchildren daily from Newton to Corfe for a return fare of 7*s.* 6*d.* (somewhat high for some six miles; possibly the contemporary report was in error and it was 7*s.* 6*d.* per week). An extra shed was built beside the locomotive shed where the Newton link line met the Norden workings, to accommodate both engine and "coach".

A curiosity of this line was that it used "stub-points"; that is, instead of having point blades, the whole rail moved sideways (*see photographs*). At the south end of Goathorn depot there was a three-way stub-point, and west of the Wareham road crossing there was a "stub-catchpoint" where only one rail was moved, to ditch any wagon which might run across the road. Normal pointwork was used on the Newton–Norden link. Another exception was at the place where trains from the Norden works to the exchange siding had to reverse, which at any rate in narrow gauge days was a normal bladed point. It seems from old maps that this "reversing siding" was in operation as early as 1880, and it may seem curious for such a clumsy arrangement to survive for 90 years; but after 1905 it may have been operationally desirable for the engine to be leading as the loaded train came down across the main road, but for the wagons to be leading when it arrived at the exchange siding, for easy release of the engine.

Some light 22 in. gauge track was used at Norden between shafts and weathering beds. Small wooden wagons, some with eyed straps for vertical lifting were in use.

The engine *Tiny* will be described in some detail, since like all Lewin engines she seems to have been unique, whereas *Thames* was of course a standard design. Her principal dimensions were: cylinders 5 in. x 9 in., boiler pressure 120 lb (later 160 lb); wheels 1 ft 9 in. diameter (later 1 ft 7 in.), wheelbase 3 ft 10 in., boiler diameter 2 ft 4 in; tank capacity 130 gallons, coal 4 cwt., weight in working order 7 tons. She was certainly rebuilt at various times, she had new cylinders in 1916, these being stamped "Dorset Iron Foundry", a firm with some connections with Lewin. At the time the author knew her best (about 1930) she had a stove-pipe chimney, large brass dome on the first boiler ring, polished bands around the cylinders, and a back weatherboard which was peculiar in that it carried two square spectacles on the outside of it, but these were not visible on the inside, and appeared to be mere decoration. The buffers were wood-faced, and the brake blocks were of wood and a cylinder lubricator was mounted in front of the chimney.

In 1938 *Tiny* got her last new boiler, from Bagnalls, and this was the one which later was mounted on one of the Pikes engines; about the same time she was fitted with a proper cab, a luxury which was customarily omitted at the time she was built, but which in less kind climates than that of Dorset became *de rigeur* much earlier.

Several building dates for *Tiny* have been suggested; one was 1868, because there was a rumour that someone had found some spare cylinders in the shed bearing this date, but the authors of the definitive history of Stephen Lewin discount this, saying the cylinders were probably from some other equipment. 1870 and 1873 have also been given (the latter to the author by her driver). Nor is it clear what she did on arrival; Mr C.N. Sykes says that she worked on the Middlebere line but was withdrawn as she "cracked the rails" — this could apply to tram-plates, in which case unflanged wheels would have been provided. Another theory was that the Middlebere line was relaid with light edge rails. Maybe she worked at Newton, opened in 1868, and she really was built in that year: the debate continues.

The wagons were short unsprung four-wheelers with three outside frame members supporting the low planked sides. The dumb buffers were set at the extreme edges of the buffer beam, forming in fact an extension of the frame to which the axle boxes were bolted. One end of the wagon could be dropped over the buffers for loading.

Roads suitable for motor traffic were now penetrating the heathland, and the role of Fayle's Tramway as the only way from Corfe to Poole Harbour was passing. It is of interest that in the late 'twenties stone was taken down the tramway to Goathorn for shipping and dumping at the "training bank" which protects the entrance to Poole Harbour between Studland and Sandbanks.

By 1937 the Eldon Siding–Newton line seemed to be little used. On a visit at Easter of that year, *Tiny* was doing its usual chores between the Norden pits and Eldon siding and *Thames* was in the newly-erected ancillary locomotive shed, together with the "coach". From here to Newton the rails showed no sign of recent use, but the portion between Goathorn Pier and Newton village was being "operated" by a trolley carrying a large barrel bearing the word "Oporto". One presumes that it carried water rather than port, and that the water was travelling from the village to the pier.

During the 1939–45 War much of the line across Newton Heath was taken up and there was no working north of the engine shed. When therefore in 1948 it was decided to regauge the line to 1 ft 11½ in., there was not a great mileage to be dealt with. To work the narrow gauge line, one steam engine and several internal combustion engined rail tractors were purchased. The steam engine was *Russell*, a 2–6–2T by Hunslet (1906) with a chequered history. She had been built for the Portmadoc, Beddgelert & South Snowdon Railway, which never got as far as opening, and she went to work for the associated North Wales Narrow Gauge Railway. When this railway became part of the Welsh Highland Railway, she was rather severely butchered to reduce her vertical clearance enough to go through the notorious Moelwyn tunnel on the Festiniog Railway, it being intended to haul through trains from Dinas Junction to Festiniog. She got through the tunnel — only just, once — but continued mainly working the Beddgelert–Dinas section, until 1936 when the line closed. A spell of five years in Dinas shed was followed by six years working for the Ministry of Supply at the Hook Norton mines in Oxfordshire, and in 1948 *Russell* came south to Norden. But there was not much life left in her. The leading axle gave trouble and she latterly worked as a 0–6–2T. In 1953 a further boiler certificate was refused, and she was purchased by the Birmingham Locomotive Club for exhibition at the Narrow Gauge Museum at Towyn in North Wales, being moved in August 1955. Ten years of standing in the open by the road bridge over Towyn Wharf station did little to improve her, and when on 15th April, 1965 she was moved yet again, to the protection of the Welsh Highland Railway Society, there was much work to be done on her if she were to steam again. This work has now been finished and *Russell* steams again.

The five internal combustion engines comprised two diesels by Orenstein & Koppel, two by Ruston, and a Motor Rail petrol engine, later replaced by a Lister-engined diesel; all were second-hand.

On the new gauge, the clay wagons were steel tippers, though one or two of the old wagons were re-gauged for special purposes. However, the use of stub-points continued and Fayles Tramway remained an interesting oddity performing a very useful job. In 1949 Fayles had amalgamated with their neighbour, Pikes, and when in 1956 the latter turned over to road transport it seemed likely that Fayles would do the same in due course.

After the War there was little left around Newton; at Goathorn the winch-gear had been thrown into the water along with a wagon, and the pier demolished. Only rubble remained around the village; a few small kilns were left, some using pieces of light flat-bottomed rail suggesting the tramway there must originally have been horse-worked. By 1969 Eldon Siding was overgrown and point-clamped; in October of that year a driver stated that the last train would cross the bridge within weeks, but in fact some working continued. In April 1970 an enthusiast noted a full train in a siding west of the main road, and two engines and 18 wagons standing on the east side lines. One year later after track-lifting had begun, an engine was working with flat wagons on the west side. All track was lifted on the east side apart from that in the old engine shed, but on the other side some 2 ft track was left; also much 22 in. track was *in situ* with tubs on it, and a cable-worked adit at the north end was still working in 1978.

One odd thing about Fayle's Tramway was the bridge across the LSWR at Norden. Clearly the Company could not have had a right of way across this piece of land when the Swanage Branch was being built, or they would have to have constructed a proper bridge. What there was in fact comprised a lattice iron arch set into unequal piers, with side girders and hand-rails which were 4 feet lower at the east side than the west side, the track-bed being built up at one end to compensate. It only makes sense if, when constructed, the gradient was down to the west, instead of, as now, down to the east.

PIKES TRAMWAY

The firm of Pike Brothers began winning clay about 1790. The Tramway is stated to have been opened about 1840; in 1866 the first-recorded engine arrived. At first the line began at a large pit to the south of Furzebrook House and ran with ruler-edge straightness to the River Frome. This was on a gentle rolling radient, ideal for gravity-working loaded and horse-working empty. The line as it later existed passed north of Furzebrook House to the Creech beds. A well-equipped depot was set up north of Furzebrook House, and to the west sidings fanned out over the clay mines.

By 1890 the track had reached Grange mine, 1¼ m. south-west of Furzebrook depot, and by 1920 Povington mine, ¾ m. further; also a branch from Furzebrook ran south-east to Greenspecks and Cotness mines. The gauge of the "main line" was 2 ft 8½ in.

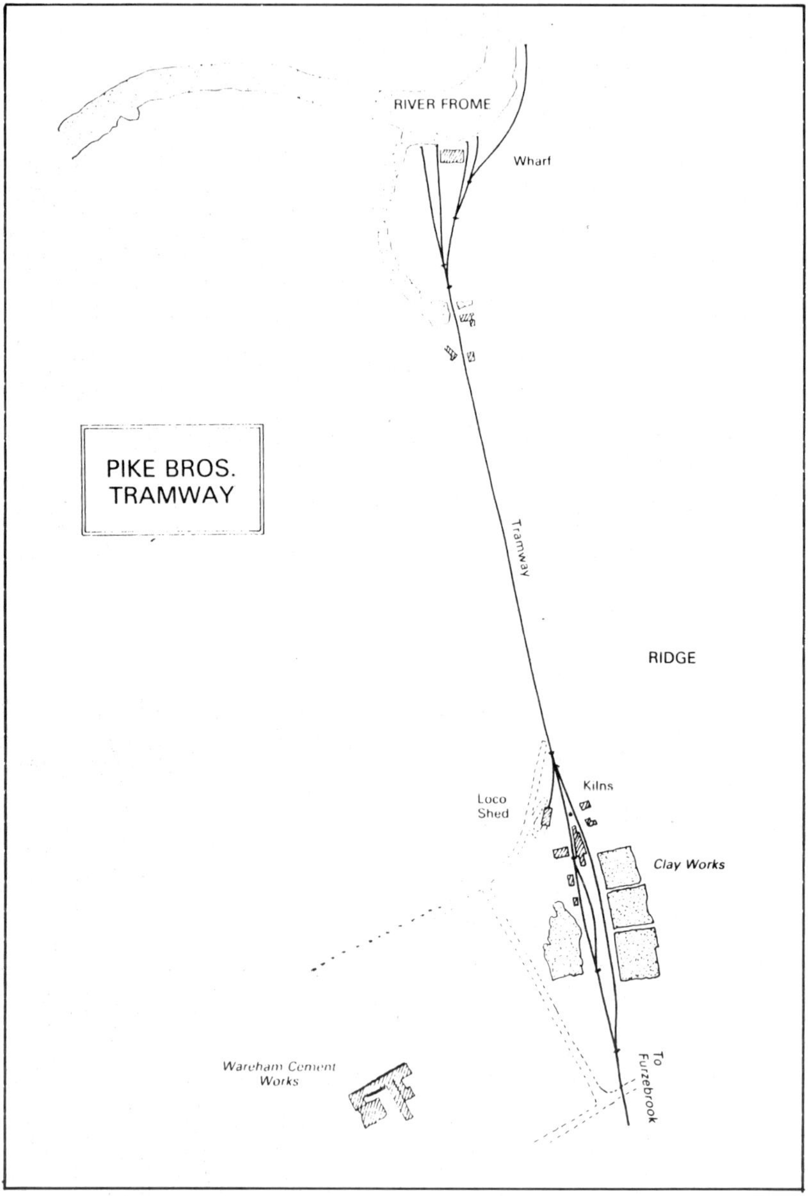

The northern end of Pike Bros. Tramway before the turn of the century.

The Newton Tramway in Goathorn Plantation photographed on 16th April, 1938.
Author

Goathorn Depot on the same date. *Author*

Goathorn Pier in 1938, showing the shute by which the clay was loaded into barges. *R.W. Kidner*

Stone plateway chair *in situ* on the same date. *Author*

Middlebere Plateway cutting and formation. 1st October, 1972. *Author*

Tiny, the Lewin engine on Fayles Tramway, 23rd August, 1932. *Author*

The two engines marshalling full and empty trains at Eldon Sidings on the same day. *Author*

From the footplate of the Lewin engine at Eldon Siding. *Author*

Typical wagons in "broad gauge" days. *Author*

The link line across Newton Heath, 1932. *Author*

Tiny crossing the SR with a load from Norden on 21st August, 1938. *Author*

Manning Wardle engine, *Thames* 1932.
Author

Ex-Welsh Highland Railway 2–6–2T *Russell* working on the newly-laid 2 ft gauge track at Eldon Siding in 1949. *Author*

Stub points at the loco shed. *Author*

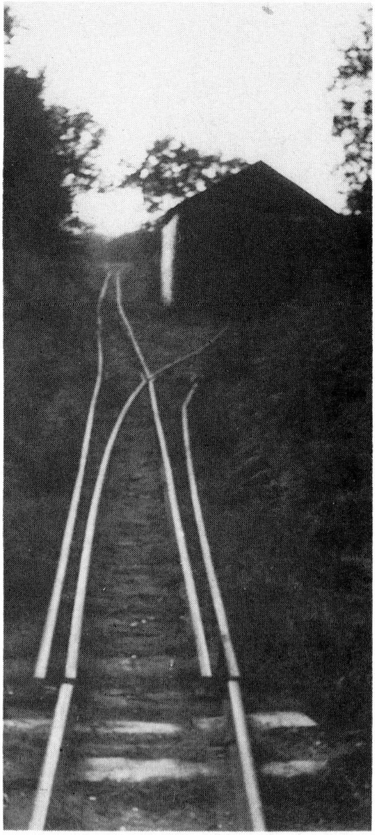

The Lewin engine on Fayle's tramway propelling a full train to Eldon Siding in 1938. *Author*

A jumble of disused stock at the Norden works of Fayles, with the main line to the pits and shafts seen on the right, 17th August, 1972. *J.R. Bonser*

The locomotive shed at Norden on 4th July, 1955. The small shed on the left had been used to house the converted wagon which formed the school train from Newton pre-war. *J.R. Bonser*

Orenstein & Koppel 0-4-0D No. 20777 of 1936 takes an empty wagon across the road near Pike's works at Furzebrook on the 4th July, 1955. *J.R. Bonser*

One of the tubs used on the 22 in. gauge tracks, at Norden in 1978. *D. Gould*

Middlebere Tramway: the northern tunnel under the Wareham Road, west face. "BF 1807" is inscribed on the keystone. *Author*

Fayles Tramway: winch-gear and a wagon in the water at the destroyed Goathorn Pier, 1977. *Author*

A cable-worked 22 in. gauge rail adit at the north end of Norden (west side) works in 1978. *Author*

Looking north from the loop at Eldon Sidings in October 1969 (Lister engine). *Author*

The reversing point on the route from Norden works to Eldon Sidings in 1969; straight ahead led to the long-disused incline. *Author*

Norden east depot in 1969. *Author*

The Lister locomotive seen on 1st October, 1969. *Author*

In the siding north of the main road at Corfe *Russell* waits (out of use) on Fayles tramway, together with a diesel wagon, while an O&K diesel engine passes on a train from the pits; 4th July, 1955. *J.R. Bonser*

Stub catch-point guarding the main road at Norden, 1969. *Author*

Diesel wagon on broad gauge frame regauged to 1 ft 11½ in. *Author*

Narrow gauge bogie wagons for bagged products, 1969 at Norden. *Author*

Junction between Works line (*left*) and Ridge line (*right*) at Furzebrook, photographed 21st August, 1938; the locomotive is *Secundus*. *Author*

Workers from Pikes going home to Ridge by gravity: 26th August, 1932. *Author*

Septimus outside the locomotive shed of the Furzebrook Tramway on 21st August, 1938. *Author*

Ridge Works and locomotive shed viewed on 26th August, 1932. *Author*

Pike Bros *Septimus* near Furzebrook in 1938. *Author*

Pike Bros *Sextus* outside the locomotive shed at Furzebrook depot in 1938. *Author*

Pike Bros riding wagon No. 48 at Ridge in 1932; the sprung buffer beam was only fitted at one end of each wagon. *Author*

At Furzebrook on 6th July, 1955; the rebuilt *Tertius* running past the small engine shed.

J.R. Bonser

Pikes tramway *Tertius* with a loaded train near Furzcbrook on 6th July, 1955.

J.R. Bonser

When the boiler of Fayle's engine *Tiny* was used to keep Pike's *Tertius* in service, the firebox would not fit between the frames and this ridiculously high boiler resulted; 6th July, 1955.

J.R. Bonser

At Ridge, just short of the wharf, was another small works and a locomotive shed (still standing). A long loop ran round the works here, with a smaller loop in the "main line" adjacent to the depot.

No doubt gravity-working was sometimes employed; even in the 'thirties the workers who lived at Ridge returned home by gravity sitting in a single braked wagon. Some of the wagons had sledge brakes (i.e. acting on the rail itself) and such powerful braking suggests gravity runs. The line to Ridge crossed the Wareham road on the level; this crossing was protected by a catch-point on the Furzebrook side and by a heavy hinged beam across the track on the other side. When the workers came down in their wagon, one would jump off, hold the catch-point lever over, double across the road to lift the beam, then jump back on the wagon without it coming to a stand — this sort of dexterity was found on many mineral railways up and down the country.

The locomotive history is not entirely clear, as the Company kept no records, but the following list is as comprehensive as records show:

Primus: a six-wheeler, probably 0–4–2WT, bought from a firm called Bellis & Seekings in 1866, and scrapped some thirty years later.

Secundus: an 0–6–0T of odd design, also bought from Bellis & Seekings in 1874, with enclosed motion and "cow-catchers" which suggest it may have been designed as a street tramway engine. It also had a marine type boiler. This engine is now preserved.

Tertius: an 0–6–0ST by Manning Wardle built in 1886 (No. 999); of their standard design, but made very non-standard at the end of its career when in 1951 the boiler of the Lewin engine from Norden was fitted to it which, having a firebox too wide to pass within the frames, was perched above them!

Quartus: said to be an 0–4–2WT by Fowler, purchased in 1889, scrapped 1934.

Quintus: a Manning Wardle 0–4–0ST (No. 1854 of 1914), which worked until 1956.

Sextus: a Peckett 0–4–2ST of 1925 (No. 1692), which also worked until 1956.

Septimus: a Peckett 0–4–2ST of 1930 (No. 1808) which differed from the above in being somewhat smaller. This engine was purchased in 1955 by the North Somerset Light Railway, which was to have run from Worle to Clevedon over the bed of the old Weston Clevedon & Portishead Railway, but which never opened. It went to Peckett's works in 1956 and was scrapped by Joseph Pugsley of Bristol early in 1962.

The main dimensions were:

	length over buffers	total wheelbase	coupled wheels
Secundus	11 ft 10 in.	6 ft 0 in.	2 ft 8 in.
Tertius	15 ft 0 in.	7 ft 6 in.	2 ft 0 in.
Quintus	16 ft 9 in.	4 ft 6 in.	2 ft 0 in.
Sextus	18 ft 10 in.	10 ft 0 in.	2 ft 6½ in.
Septimus	17 ft 8 in.	9 ft 6 in.	2 ft 3 in.

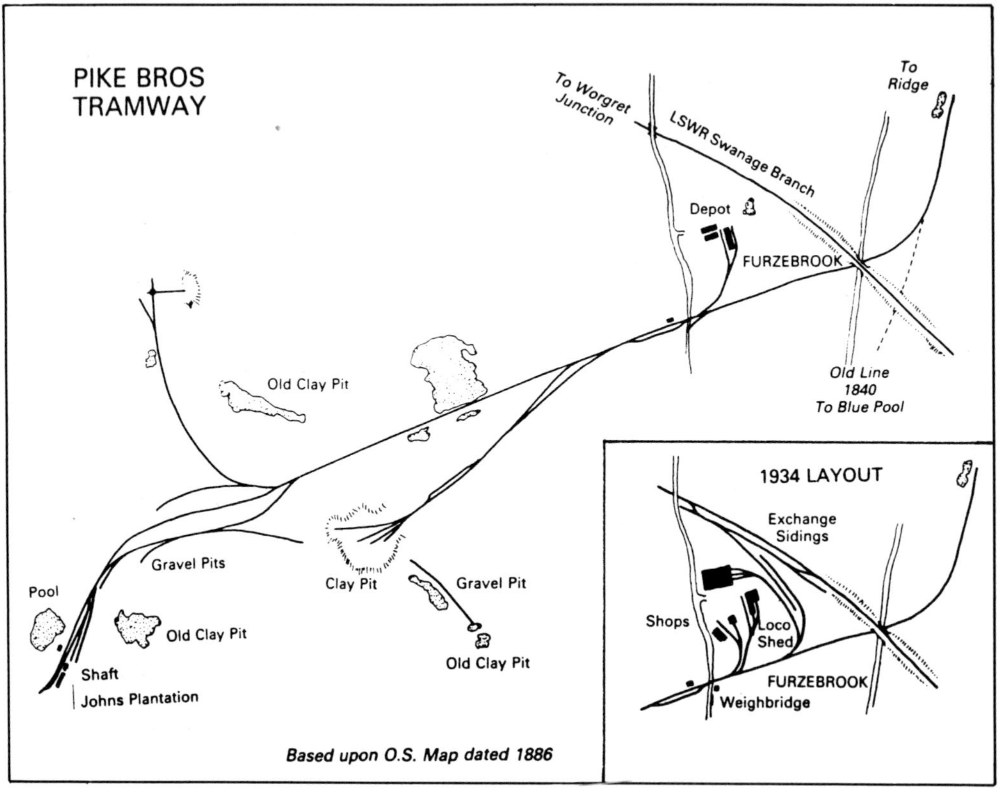

Based upon O.S. Map dated 1886

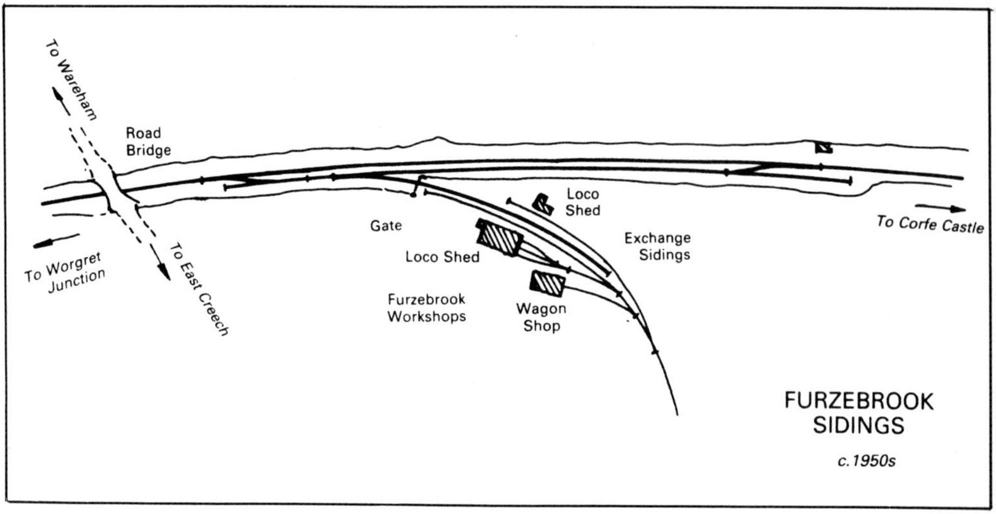

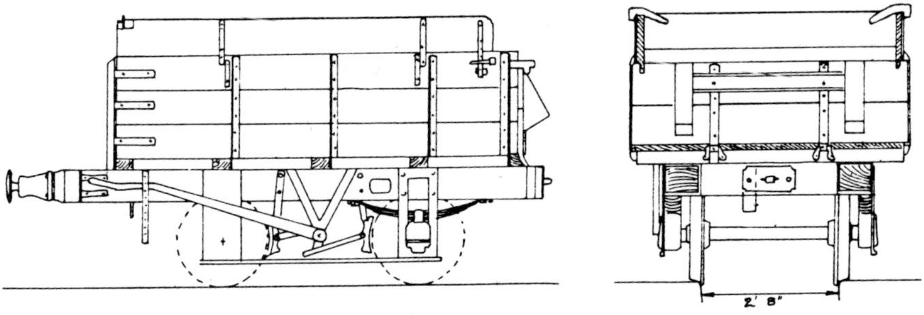

Furzebrook Tramway (*Pike Bros. Fayle & Co. Ltd.*) 4-ton end door wagon (*in original condition*). Built by Cambrian Wagon Co.
Drawn from information by Cambrian Wagon Co. and photographic evidence.

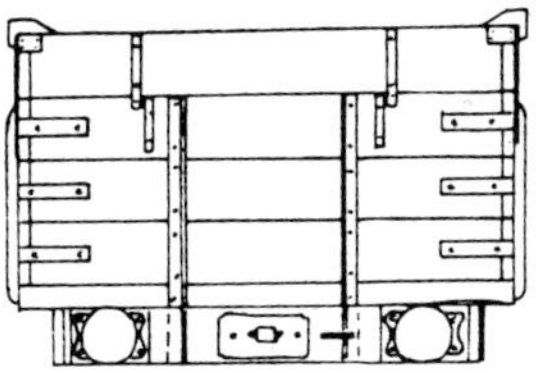

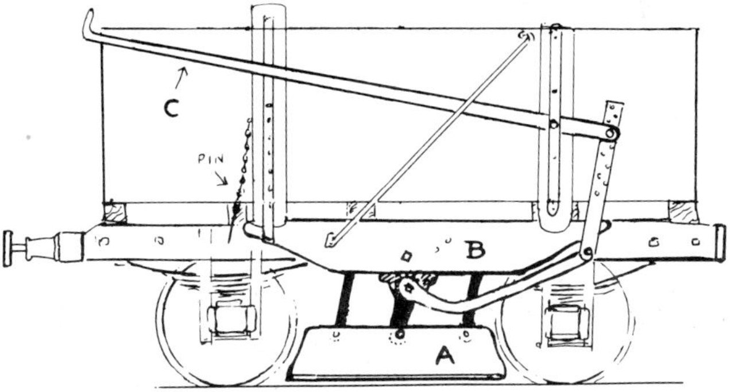

Pike's Tramway: operating method of sledge brake; pushing down the lever C causes sledge A to bear on the running rail, acting against thrust beam B fixed to side of wagon.

In 1951 a Simplex diesel locomotive was bought, second-hand. The diesel-oil tank which was set up on a stage at Furzebrook for fuel later saw service mounted on a converted "broad gauge" wagon on the Norden line.

In 1955 road transport began taking over, the first section to go (early in the year) being Cotness. Greenspecks followed in the summer, and the last run to Creech Grange was on 30th July, 1956. The weathering beds were rail-worked for a further year. The Povington mine was on Army land and may have been disused before the rest ot the branch.

In the early days T-bulb rail in chairs was used: some remain as fence-posts today. Later flat-bottom track was used, and in the mines themselves some 22 in. gauge light bridge rail have been found.

The wagon stock has been fully described in a booklet published by the Narrow Gauge Railway Society; it comprised end-tipping, side-tipping, and fixed wagons of various kinds, some built by the Cambrian Wagon Co. and some locally. One of the sledge-braked wagons was rescued and is now in the Towyn Narrow Gauge Museum. There was one odd thing about the wagons: the buffer beams were sprung at one end only, by a leaf-spring set transversely between beam and frame. This did not apply to the end-tippers which were mostly used in the mine areas; these had only dumb buffers.

Although the wagons appeared to be of great antiquity, one enthusiast who came to mourn at the pyre of burning wagon stock in November 1957 reported that some had makers' plates as late as 1936.

One curious feature of the end-tippers was that the body overhung the frame at one end, with a space at the other end; thus if all were lined up with the tipping end facing the weathering beds, all was fine; but the engines had to work with flat-wagons attached (as in the photograph of *Septimus* at the engine sheds) so as to be able to couple to a tipping end if required.

Some of the 1 ft 10 in. gauge tramways in the clay workings continued in use after the Furzebrook Tramway itself closed: in 1968 the Pike's undertaking was purchased by English China Clays.

Secundus is preserved in the Newhall Street Museum in Birmingham: a very interesting though hardly typical specimen of Victorian engineering. How many of her oddities, which include outside Stephensons valve gear, were due to a rebuilding by Stephen Lewin in 1880 is hard to say.

The BR siding at Furzebrook works is still in use (*see later*).

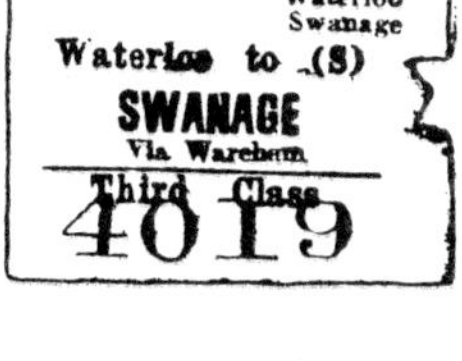

Chapter Four

Other Tramways

THE SWANAGE PIER TRAMWAY

This line was one third of a mile long, and ran from the old pier to stone stores and a coal depot along the seafront. The Tramway was designed by Capt. W. Moorsom, Chief Engineer of the Southampton & Dorchester Railway, and seems to have been opened in 1858, notwithstanding a print of it published in Poole dated August 1856. The gauge was 2ft 6in. The pier was built to obviate the rather dangerous practice of horse-carts going out into the waters of the bay to load stone onto barges. There was a plan to build a tramway from the pier to the stone quarries at Langton Matravers. There can be little doubt that at any rate until the coming of the LSWR there was considerable traffic in stone up the pier line, but it does not seem that mechanical traction was used. At the time the author first knew it (in the 'thirties) only one wagon remained, apparently used for carrying fish, but it is on record that quite a number of stone wagons were in use earlier, hauled by horse-teams. The present steamer pier is a later structure, added in 1896. According to the Dorset Year Book, traffic from the pier comprised stone outwards and timber inwards by day, and coal by night.

The layout has changed at various times. About 1885 a large fish store was built half-way down the Tramway, with a short branch entering it. It seems that bodies fished from the sea (and there were many wrecks hereabouts) also went down the Tramway and part of the fish store was used as a morgue. An extant photograph of stone wagons standing in the short siding outside the fish store shows short four-wheeled wagons with outside journals, and no sides.

The Tramway is still in place for about a hundred yards from the fish store towards the pier; most is in flat-bottom rails of 30 ft lengths, but part is of tramway groove-rail. The pier itself has been allowed to decay, though still standing.

THE BROWNSEA ISLAND TRAMWAY

In the early 1850s a somewhat fraudulent character, Col W.P. Waugh, discovered clay on Brownsea Island. He borrowed half-a-million pounds and proceeded to lay down equipment to work it: two potteries, a 200 ft drying shed, a substantial pier, coal pens, reservoir, and a narrow gauge railway serving these points and the mines which were mostly in the centre of the island. A company was formed, the Branksea Clay & Pottery Co. However, the seam turned out to be very thin; the Colonel had an extravagant life-style and profits were meagre. In 1857 he was made bankrupt, but working was carried on by the Trustees. In 1873 there were 240 people living on the Island, but from 1877 the industry was run down. With some 2 miles of track, the railway might have used steam power, but there is no firm evidence to suggest this.

THE KIMMERIDGE SHALE TRAMWAYS

Interest in the shale deposits of Kimmeridge has lasted for nearly 2000 years, and an oil rig is now pumping oil on almost the same spot as a Roman

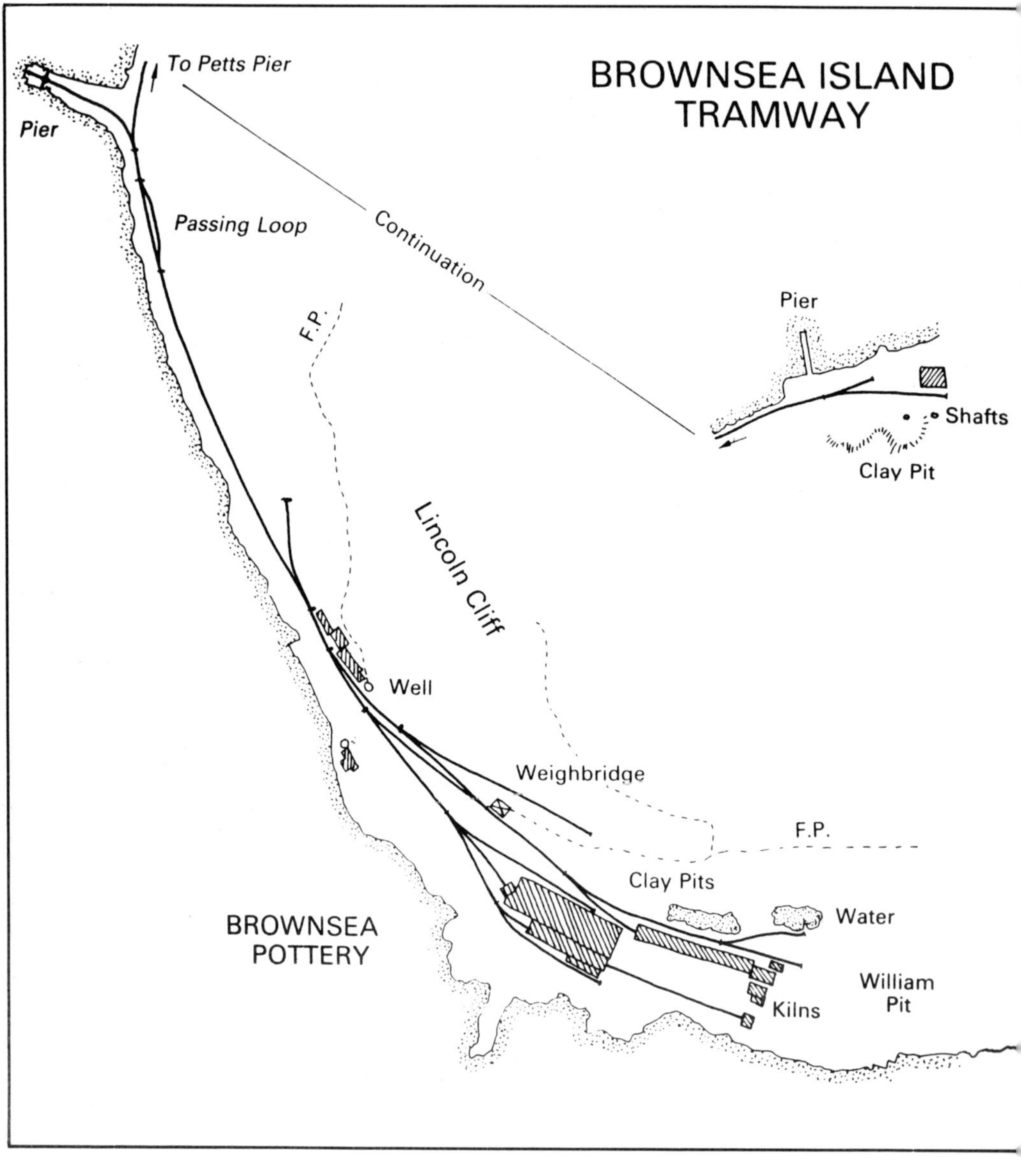
BROWNSEA ISLAND TRAMWAY
To Petts Pier
Pier
Passing Loop
Continuation
F.P.
Pier
Shafts
Clay Pit
Lincoln Cliff
Well
Weighbridge
F.P.
Clay Pits
Water
BROWNSEA POTTERY
William Pit
Kilns

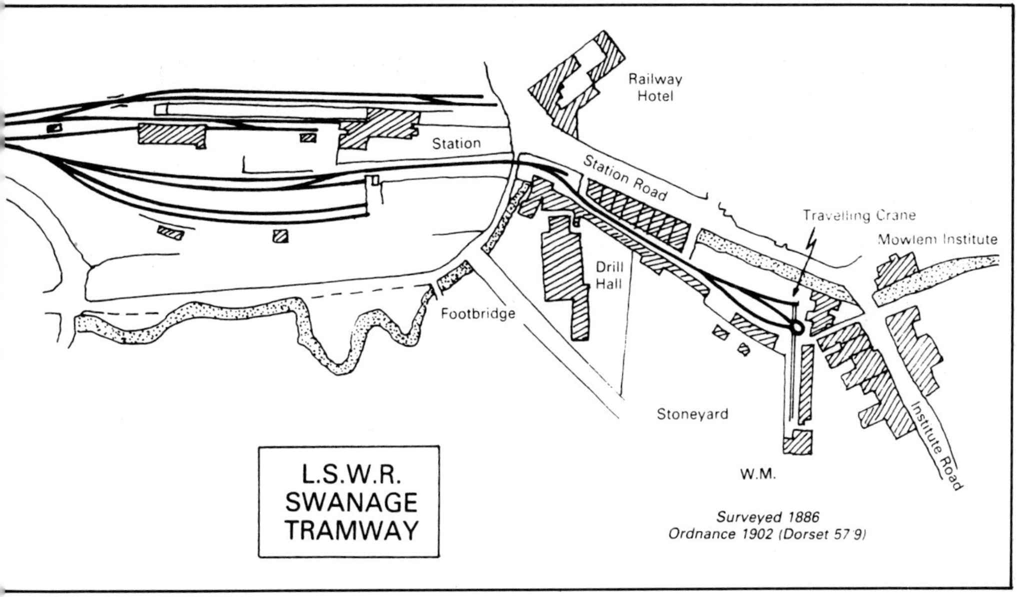
Railway
Hotel
Station
Station Road
Travelling Crane
Mowlem Institute
Drill
Hall
Footbridge
Stoneyard
Institute Road
W.M.
L.S.W.R.
SWANAGE
TRAMWAY
Surveyed 1886
Ordnance 1902 (Dorset 57 9)

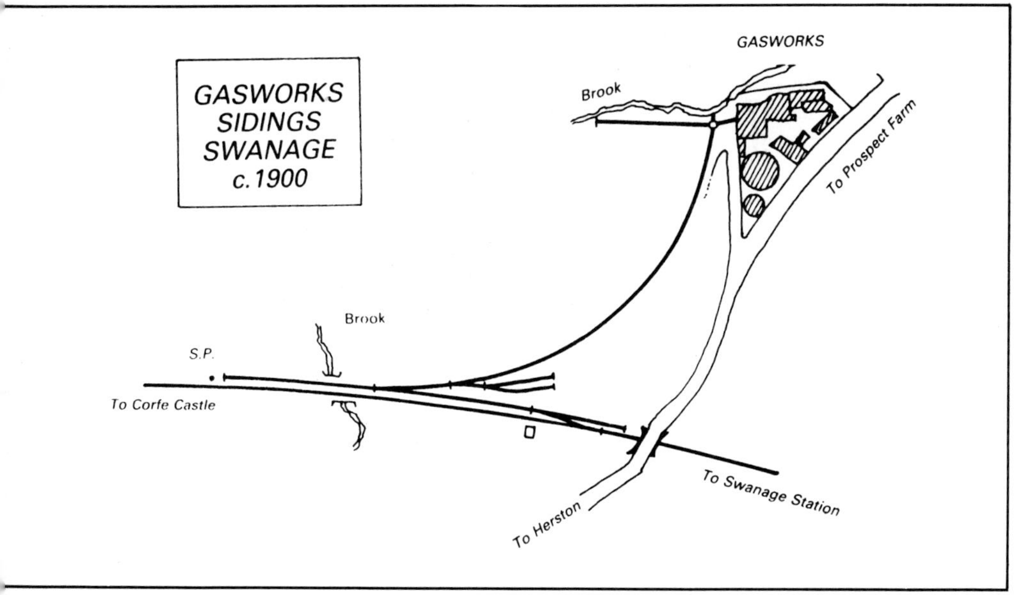
GASWORKS
GASWORKS
SIDINGS
SWANAGE
c.1900
Brook
To Prospect Farm
Brook
S.P.
To Corfe Castle
To Herston
To Swanage Station

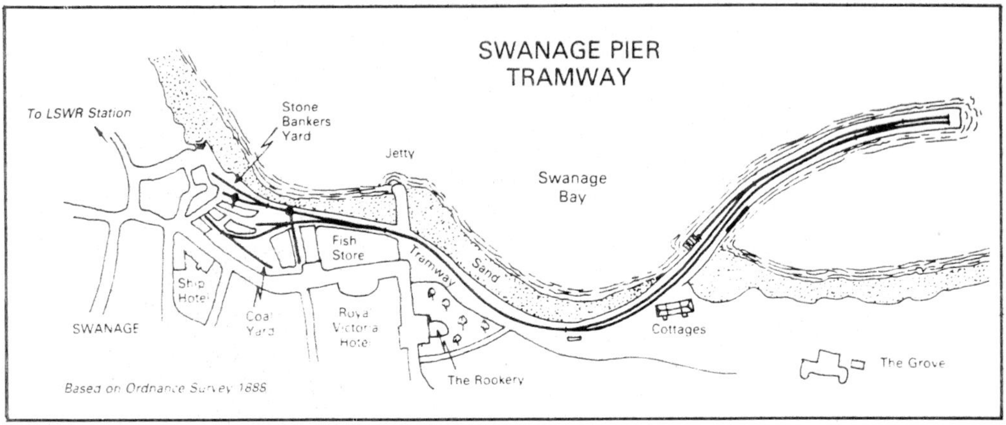
SWANAGE PIER
TRAMWAY
To LSWR Station
Stone
Bankers
Yard
Jetty
Swanage
Bay
Fish
Store
Tramway
Sand
Ship
Hotel
Coal
Yard
Royal
Victoria
Hotel
SWANAGE
Cottages
The Rookery
The Grove
Based on Ordnance Survey 1886

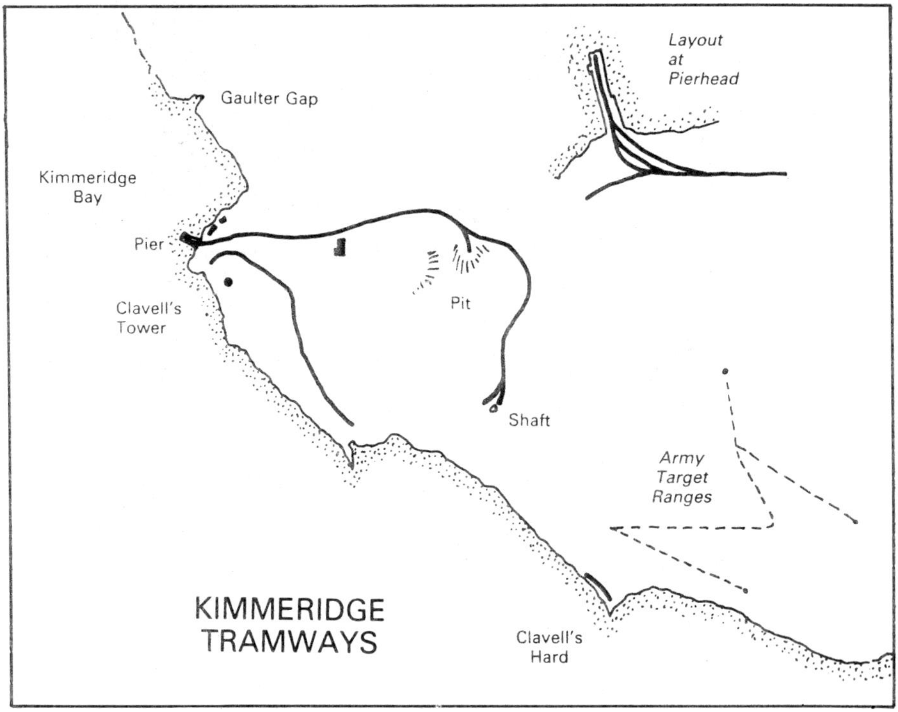
Layout
at
Pierhead
Gaulter Gap
Kimmeridge
Bay
Pier
Clavell's
Tower
Pit
Shaft
Army
Target
Ranges
KIMMERIDGE
TRAMWAYS
Clavell's
Hard

A print published in Poole on 1st August, 1856 showing the Swanage Pier tramway; a train of ten stone wagons is shown approaching the small kiosk. *Author*

Halfway along the Swanage Pier Tramway about 1900; stone yard right of centre beyond jetty. *Frith, Reigate*

The Piers at Swanage around 1904, with two wagons on the tramway pier.
Stevens, Poole

The Swanage Pier Tramway seen in May 1973 (near former fish store). *Author*

The early Kimmeridge tramway passing to the south-east of Clavell's Tower, May 1973. *Author*

Possible wagon wheel from early tramway; at Kimmeridge May 1973. *Author*

shale mine. The first use of railways here was probably about 1848, when an Act was obtained for "constructing railways, inclines, causeways etc.", and the earliest line seems to be a short one from the cliffs at Yellow Ledge round the south side of the hill which is topped by Clavell's Tower (built 1837) to a level above the east shore of the Bay. The line falls at a gradient of 1 in 40; nothing is known about it, but two lengths of heavy bridge-section rail lying in a gully at Gaulters Gap in the centre of the bay may have come from it, as also may a railway wheel being used as an anchor at the fishing "hard". This is 18 in. in diameter, cast-iron, with eight holes very similar to the wheels of wagons used to build the Plymouth breakwater about 1820. It has a pronounced flange on the inside and a small flange on the outside, and thus could not have been used on a railway having points of the normal type; this early Kimmeridge line is unlikely to have had any points and might have used only one wagon.

A longer line ran from a wooden pier at the east end of the Bay past the coastguard cottages and then east to a shale pit (map ref. 916782) near the "D Plantation" on the Smedmore Estate about half a mile from the Bay, and then turning south, another half-mile to a shaft (map ref. 915787). A date of 1858 has been assigned to this, and it has been tied in with a company which processed shale oil at Wareham to supply lighting for Paris; but the large quantities of rail from it which still can be found, as a fence around Clavell's Tower, and on the "hards" at the Bay, suggest a later date. They are all flat-bottom, ranging in weights from about 15 lb/yd. as might be found in workings, to some 30 lb/yd. suited to the "main line", and seem unlikely to have been rolled so early. The route at the Bay end has been covered by tracks for holidaymakers, but inland the formation can be found if searched for carefully. A painting by Mr C. D'Oyly in Smedmore House shows the wooden pier *in situ* about 1900, and Sir Frederick Treves writing in 1906 stated that there were still wagons to be seen on the pier. Portions of four main supports for the pier can still be seen amongst the rocks.

The Bituminous Shale Company, who processed shale at Weymouth were active around 1850 and probably used the early line. Messrs Ferguson & Muschamp took over and moved the processing to Wareham, being only interested in turning the shale into fertiliser; there are no details of their method of transport. A French company under Marshall Pelissier was working the shale in the eighteen-seventies and is likely to have built the second line and the pier. The company was wound up in 1872, and no details have been found of any later use of the Tramway.

A third tramway is thought to have existed on a working level half-way down the shale cliffs at Clavell's Hard, east of Kimmeridge Bay; there are tunnels here, and records of rails and iron ladders, together with iron sockets in the "hard" itself which could have been used in connection with a telpher or incline to enable the shale to be loaded into barges.

ARMY TARGET RAILWAYS

During the 1939–45 War the Army laid down some railways for practising firing at moving targets, above the cliffs near Clavell's Hard; this was an ideal position as the shells would fall in the sea; a similar railway was laid on Seaford Head in Sussex. After the War the rails were lifted and the formation ploughed in.

THE COCKNOWLE TRAMWAY

A balance-worked railway incline is said to have been here a hundred years ago; the site is almost certainly the formation of about 150 yds in length, barely perceptible amongst the gorse, which runs straight up the north face of Knowle Hill opposite a recently-opened chalk pit (map ref. 938821). What seems to be another railway, 200 yards long, runs from the foot of this incline on an easier gradient, to the west. This is slightly curved and could not have been balance-worked, but would have needed some sort of cable assistance. The object was the marl dug from pits at the top of the hill, and it was transported in carts to Ridge, where it was processed by the Wareham Cement Company at a premise close to that of Pike Brothers.

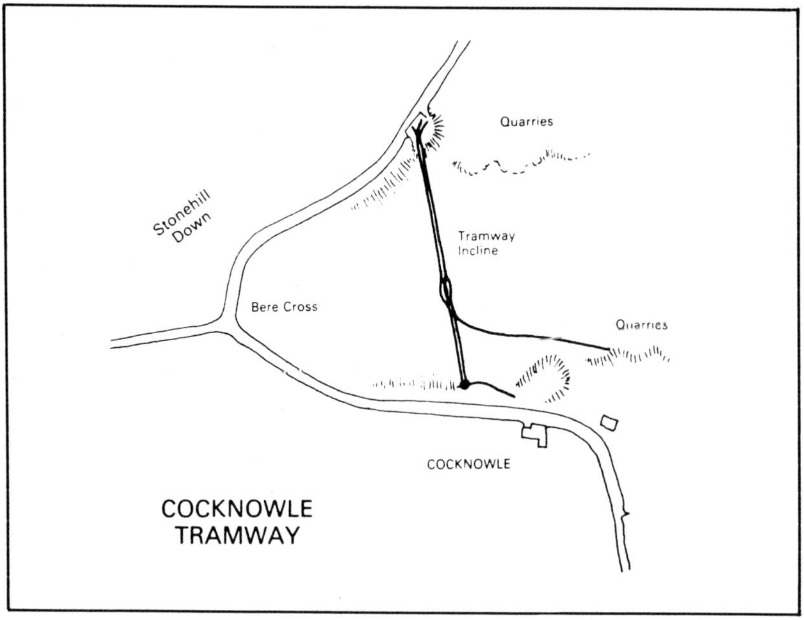

Chapter Five

The Swanage Railway: 1973–1988

The idea of re-opening the Swanage Branch was being actively canvassed even before it closed. British Rail moved with their usual indecent haste to take up the track, and mounting protest in August 1972, only stopped them, by which time the only stretch which could be saved was about a quarter of a mile southwards from Furzebrook. The future did not look good; British Rail, giving the no doubt accurate picture that the last thing they wanted at Swanage was a railway, proposed to develop the station site. However, in the end it was sold to the Town Council for a car park — though the Council was not totally opposed to the idea of a railway. Much patient lobbying ensued; in 1975 a local vote showed 83 percent of the townspeople in favour of the railway. In the same year the Dorset County Council, which now owned the trackbed apart from the Swanage station area agreed to lease it to the new railway; but such was the unrealistic thinking that this was made contingent upon the Swanage Railway being able to run a service of at least four trains per day to Wareham by January 1980.

Plans continued to be made; it was decided to have stations at Herston, Harmans Cross, Corfe Castle, The Blue Pool (from where track was still in) and Worgret (just before the junction). The clay traffic was of course still running, though the oil problem had not yet arisen. BR were prepared to consider their own staff taking trains on from Blue Pool to Wareham. However, as previously stated, there was still another hurdle; the authorities wanted to use the track-bed through the gap at Corfe for a by-pass.

Some stock was purchased: the 4-wheel Planet diesel locomotive *Beryl* from Hamworthy coal wharf, an ex-SR class 4, 2–6–4T No. 80078, and an ex-SR Utility Van No. 1234, later DS3065. These were placed in the old goods shed. The body of an ex-LSWR 6-wheel 6 compartment 3rd class carriage No. 695 was obtained, placed on the cattle dock at Corfe, and restored to the LSW pinkish livery. At Swanage a short length of track was laid alongside the station platform but shortly afterwards moved to the old coaling platform and scenery van 4594 put there for painting. In 1977 some further track was lifted from Eldon Sidings, where it had been the property of the clay company and not of BR, and in 1978 an ex-GWR hand-crane which had been purchased two years earlier arrived. By now a Bulleid corridor coach No. 4365 had come from WD Long Marston (Army 5200), and was placed in the goods shed for extensive restoration. More motive power came: ex-GWR 0–6–2T No. 6695 (1928), a Fowler 1957 0–4–0DM *May* from the Ministry of Works in Manchester, and *Richard Trevithick*, an 0–4–0ST Barclay (1954) from Godrington Power Station. There was a small victory on 28th August, 1979 when it was possible to announce a diesel-hauled train service, but only as far as the Northbrook Road bridge.

During 1979 about ¼ m. of further track was laid towards the site of the gasworks siding. However this was mostly ex-London Transport material and not very impressive. Viewing the GWR 0–6–2T which was minus every brass fitting, and the 2–6–4T which was in three pieces and awaiting a pair of very expensive replacement connecting rods (BR had cut the old ones in

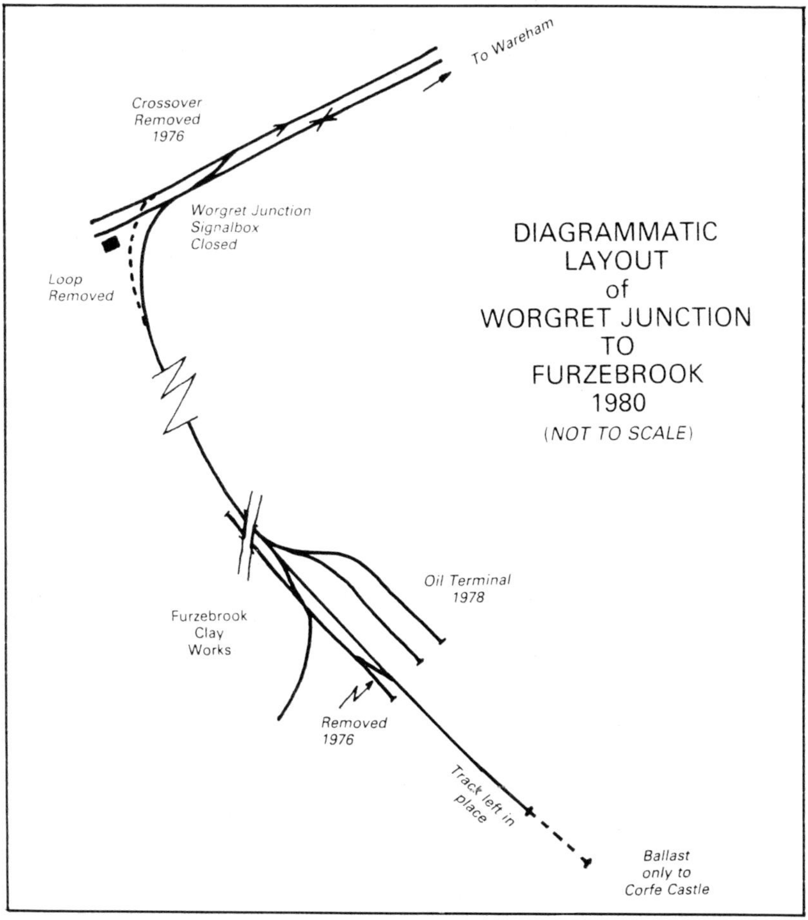

half), most observers felt that success was very unlikely. However, perseverance had its reward. For the 1982 season a service was run for 1 mile to Herston; Harmans Cross (3 m) was forecast for 1983, but this could not be achieved, partly due to the necessity to rebuild Knitson bridge. More rolling stock arrived; perhaps one of the most noteworthy was de-motored "Brighton Belle" Pullman motor-coach No. S288, transferred in 1982 to Swanage from the Stour Valley Preservation Society.

Nine trains per day were provided in 1987, and the Halt at Harmans Cross was completed; motive power assistance was provided by ex-LSWR 'B4' 0–4–0T *Normandy* borrowed from the Bluebell Line. Confirmation of the Light Railway Order was also received. The Railway's literature makes much of the fact that it is not just another railway from "nowhere to nowhere" — and everyone knows some of those! If it can overcome the financial and political problems it can become a railway offering easy access to London as well as lovely scenery for the tourist, and probably also some freight traffic over part at least of its mileage. The difficulties are still immense, but the operating Company and its supporting bodies, the Swanage–Wareham Group and the Southern Steam Trust, fully intend to succeed.

FURZEBROOK BRANCH

What is left of the Swanage branch now falls into two distinct sections: the "Swanage Railway" gradually extending from Swanage northwards, and the original line from Worgret Junction as far as Furzebrook, used by the clay and oil trains and still in British Rail ownership.

As mentioned earlier, about a quarter of a mile of track still exists extending south from the run-round at Furzebrook, though this has no use at present. From the buffer stop at the end of this section the formation continues undamaged and Corfe Castle station is in good care.

Regular trains for the products of English China Clay at Furzebrook continued after the passenger service ceased, usually 4 or 5 wagons per day, which were dealt with at Wareham, the empty wagons also being stabled there.

The Worgret Junction cross-over was taken out in November 1976, and thereafter the loaded trains ran "wrong line" into Wareham, fitted in between two down trains. Meanwhile construction of a new oil terminal on the east side of Furzebrook sidings to handle the products of the Wytch Heath wells was continuing, and a double siding was laid to serve this, joining the "main line" under the bridge by Furzebrook Works. The first oil train ran on 15th December, 1978, and soon after the oil terminal on the up side at Wareham sidings was abandoned, later being lifted to allow the building of a new road overbridge on the site.

The Queen spent the night of 23rd March, 1979 in her Royal Train on the branch, before a state visit to Poole the next day; locomotive No. 47 539 handled the train, which passed through Wareham at about 1 am after normal service had ceased.

CONCLUSION

The Isle of Purbeck has seen railways of at least eight differing gauges, possibly more. It is full of interesting relics — which incidentally include two notable coach bodies unconnected with the branch, one of which was rightly being restored to SECR lake livery when last seen. In time many will disappear, with the development of the heaths for other purposes: there is always the fear that once again someone will try to drive a four-lane motorway through the Corfe gap and thus cut the Swanage Railway off from its natural outlet. But there is hope that as the years pass some veneration will arise for such things as Benjamin Fayle's tunnels. Does the pier tramway still lie in its concrete bed because it has been forgotten, or because Swanage values it? What problems will arise when the Swanage Railway finally runs into the south end of the oil terminal and on to Wareham? It is an exciting future — or can be; and that is a better ending than many railway histories can claim.

Appendix One

Swanage Railway: Stock List as at 15th March, 1987

LOCOMOTIVES

Main Line Diesel

1958 BR 350 hp diesel shunter No. D3591 (08 476).

Main Line (Steam)

1928 GWR 0–6–2T No. 6695.
1948 SR "Merchant Navy" class 4–6–2 No. 35022 *Holland–America Line.*
1954 BR Standard class '4MT' 2–6–2T No. 80078.
1954 BR Standard class '4MT' 2–6–4T No. 80104.
1905 LSWR class 'M7' 0–4–4T No. 30053.

Industrial (Steam)

1931 Hunslet 0–6–0ST No. 1690 *Cunarder.*
1938 Hawthorn Leslie 0–6–0ST No. 21.
1954 Peckett class 'OQ' 0–6–0ST No. 2150.

Industrial (Petrol/Diesel)

1937 Hibberd "Planet" 4WPM *Beryl.*
1957 Fowler 0–4–0DM No. 4210132 *May.*

CARRIAGES & WAGONS

Passenger/Parcels Rolling Stock.

1885 London & South Western Railway — three coach bodies.
1916 L&NW Rly Covered Carriage Truck No. 4183.

1925	LMS Third Open Coach No. 7868.
1930	SR Third Open Coach No. 1381.
1931	Pullman *Brighton Belle* Driving Motor Coach No. 88 (later 288).
1931	SR Brake Compartment Third Coach No. 2768.
1936	SR Luggage Van No. 1234.
1938	Bogie Guard's Van No. 385.
1938	SR Scenery Van No. 4594.
1938	LMS 6-Wheel Stove Van No. M32990.
1941	SR Passenger Brake Van.
1947	SR Brake Semi-Open Third Coach No. 4365.
1947	SR Brake Semi-Open Third Coach No. 4366.
1947	SR Corridor Composite Coach No. 5761.
1951	BR Corridor Second Class No. 24127 (Not at Swanage).
1956	BR Tourist Second Open Coach No. 4416.
1956	BR Tourist Second Open Coach No. 4349.
1957	BR Brake Second No. 35059 (Not at Swanage).
1957	BR Corridor Second No. 18424.
1957	BR Unclassified Restaurant Coach No. 1908.
1958	BR Brake Composite No. 21205.
1958	BR Second Class Sleeping Coach No. 2564.

Goods/Departmental Rolling Stock.

1919	GWR 6-Plank Open Wagon No. 97286.
1928	SR "Lowmac" No. ADS61095.
1937	ICI Liquid Ammonia Tank Wagon No. 62WW.
1942	Ministry of Supply 25 Ton Goods Brake Van No. WD49012.
1947	SR 15 Ton Diesel Electric Crane No. 1.
1947	SR "Walrus" Ballast Hopper Wagon No. DS 62070.
1955	Wickham PW Motorised Trolley No. 9024.
1955	BR 6-Plank Open Wagon (ex-PBA 63012).
1958	M.o.S. Box Van No. WD47711.
1958	M.o.S. Box Van No. WD47726.
?	PBA (ex-LMS) 8-Plank Open Wagon No. PBA58202.
?	PBA (ex-LSW) 8-Plank Open Wagon No. PBA59342.
?	PBA (ex-LSW) 5-Plank Open Wagon No. PBA59355.
?	PBA (ex-LSW) 5-Plank Open Wagon No. PBA60001.
?	PBA (ex-GWR) 6-Plank Open Wagon No. PBA61002.
?	PBA Flat Wagon No. PBA121.
?	PBA Flat Wagon No. PBA128.
?	PBA Flat Wagon No. PBA131.
1924	GWR "Mink" Box Van (Body Only) No. W104670.
1953	BR "Vanfit" Box Van (Body Only) No. B755653.
1950	LMS 20 Ton Goods Brake Van No. DB 950194.
?	Crane Runner No. ADE 320682.

Appendix Two

The Purbeck Line

SWANAGE RAILWAY

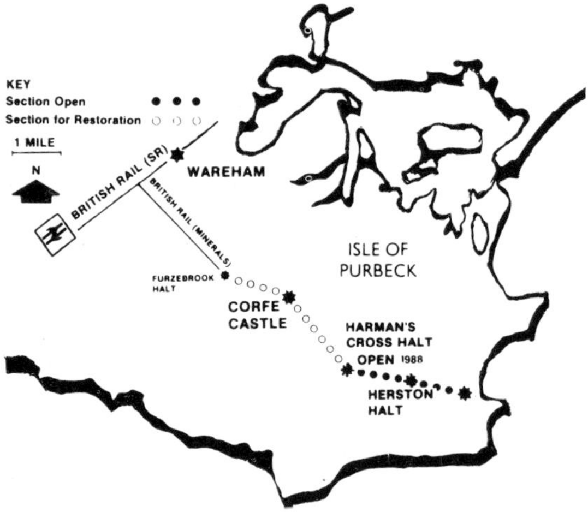

HOW TO GET TO SWANAGE

By Train — BR to Wareham and then bus hourly.

By Bus — Wilts & Dorset services: from Wareham and Bournemouth/Poole.

By Car — Take the A351 from Wareham or the road from Bournemouth/Poole via the Toll Ferry at Sandbanks.

NB: — There is no car access at Herston Halt or New Barn Bridge. There is a public car parking at Swanage Station and also at Victoria Avenue. Coaches may put down and pick up passengers at Swanage Station, but must park at Victoria Avenue Car Park.

THE PURBECK LINE